FUERTEVENTURA

Raimundo Rodríguez
Rafael Paredes

RAIS Ediciones

Title: FUERTEVENTURA
Texts: Rafael Paredes and Raimundo Rodríguez
With contributions from:
Sofía Menéndez
Miguel Pizarro
Eduardo Rodríguez
Rosa Luján
Carmen Román
Rosario Cerdeña
Lidia Soto
Rosa Guerrero
Photography: Raimundo Rodríguez
Miguel Pizarro (fish)
Aerial photography: Grafcán S.A. http://www.grafcan.com
Illustrations: José Alfonso Madrid
Translation: Martin Kreutzer (German)
Carl Mac Gabhann (English)
Photomechanics, printing and binding: Tecnographic S.L.
First edition: July 1993
Second edition: April 1994
Third edition: August 1996
Fourth edition: December 1997
Fifth edition: January 2002
Depósito Legal: SE-1125-93
Published and distributed by RAI Ediciones
C/ Manuel Sánchez Evora, 5
35630 Antigua. Fuerteventura. España
928 87 83 09 - 608 15 84 51
e-mail: rai.mundo@jet.es
http://web. jet.es/rai.mundo

CONTENTS

Introduction **5**

Geological Origin **6**

The Island's Relief **12**
Mountains. Volcanoes. Malpaís. Plains. Jables. Beaches. Coasts and Seabed

Natural Resources **30**
Climate (**30**). Settlement and Development (**31**). Flora (**32**). Fauna (**36**).
Protected Natural Areas (**44**).

History and Culture **52**
Prehistory (**52**). The Conquest (**54**). Modern History (**59**). The Economy (**61**).
Customs (**66**). Fiestas (**69**). Arts and Crafts (**71**). Architecture (**74**). Food (**84**).

Sports **86**
Fishing (**86**). Hunting (**86**). Diving (**88**). Windsurfing (**90**).

Municipalities **91**
Puerto del Rosario (**91**)
La Oliva (**92**)
Antigua (**93**)
Tuineje (**94**)
Betancuria (**95**)
Pájara (**96**)

Bibliography **98**

Touring routes (by car and on foot) **99**
Length, duration, difficulty, access and description
Corralejo - El Cotillo (100)
Lajares - Corralejo (102)
El Cotillo - Los Molinos (106)
Puerto Ventura - Villaverde - El Cotillo (108)
Salinas de Carmen - Pozo Negro - Casa de Jacomar (110)
The Betancuria Massif (112)
The Malpaíses (114)
Istmo de la Pared (116)
Pico de la Zarza (118)
The Jandía Massif (122)
Isla de Lobos (124)
The Windmill Route (126)

INTRODUCTION

Fuerteventura is much more than a sun-and-sand holiday resort. And the visitor who really wants to discover the island's soul could do nothing better than follow in the steps of the author of this guide. It is one sure way of discovering for yourself the depth and beauty of Fuerteventura's landscape, natural treasures and inhabitants.

Opening the cover of this book is like opening a door to this marvellous island of "Maxorata". The key lies in the author's sensitivity, his love for Fuerteventura, and not least, his deep personal experience - the fruit of years spent exploring its ravines, its beaches, villages and history. He has truly fallen into the trap of its magic.

When the sun sets, dear reader, please don't go back to your hotel room. Instead go up into the heights and take in the living museum of colour before your eyes. Discover its peace and tranquillity, which are the keys to Maxorata, although increasingly under threat from the onslaught of mass tourism.

Fuerteventura is, without a doubt, the most unspoilt of the Canary Islands. Its wealth is, today, boundlessly superior to all the monuments ever purpose-built by humans to attract tourists.

Let yourself be trapped by the African beauty of the island's hidden corners - just check the recommended routes - its arid majesty and the oldest volcanoes of the entire Archipelago.

Immerse yourself and swim in the blue waters of some of the beaches described by the author, for a truly unique experience - therapy to rinse your mind and body of the after-effects of today's city life.

After a long day spent strolling or on the beach, there is no better way to satisfy your healthy appetite than the local food. You can't leave without having sampled "Majorero" cheese, the island's tomatoes, gofio, a fish soup, or grilled parrot fish, among many other culinary delights.

Biodiversity is the order of the day, and not just with regard to the menu and the natural surroundings, but in everything. Try to experience the island's culture, its craftwork, the aboriginal traditions which still live on, for example - the Cofete "apañada" or livestock round-up.

In short, Fuerteventura is a truly unique island - a world in itself - a place to discover the new and broaden horizons.

When you leave, dear visitor, you will probably feel a tinge of sadness, and this guide should serve as a happy reminder of your time here. An illustrious thinker exiled on the island at the start of the 20th century wrote "... For me, Fuerteventura, was an oasis, an oasis in which my spirit imbibed the reviving waters and from which I returned refreshed and determined to continue my journey through the desert of civilisation...". Following this guidebook will probably have a similar effect on the reader of today.

Sofía Menéndez

GEOLOGICAL ORIGIN

Fuerteventura is one of the Canary Islands, a volcanic archipelago situated in the eastern Atlantic, northwest of the African continent.

There are at present some seven theories concerning the formation of the Archipelago; of these the most complex and recent holds that the islands were formed through a dual process of underwater eruptions and the thrusting up of large sections of oceanic crust. Detention of the African Plate led to compression of the Mid-Atlantic Ridge. The flow of lava occurred along directionally dominant fissures which determined the shape and structure of the islands, their alignment, and the general configuration of their dikes.

Volcanic activity in the area is not, nor has it ever been, continuous. Distinct eruptive periods or volcanic cycles occurred which, in Fuerteventura, lasted from the Miocene Era until a few thousand years ago, although the island took its present shape during the Pliocene-Pleistocene period. Regarding the present make-up of Fuerteventura, we can establish a primary distinction between materials forming the island's base (known as the Basal Complex), and the basaltic material originated during periods of sub-aerial volcanic activity.

The Basal Complex, which predominates in the western-central part of the island, comprises a collection of sedimentary plutonic, subvolcanic and submarine-volcanic rocks. The sedimentary materials derive from the oceanic crust, meaning they could be as much as 100 million years old (Cretaceous Era), vastly predating the magmatic Canary epochs and coinciding with the origins of the Atlantic itself.

The great age of the Basal Complex means it has been extensively exposed to the forces of erosion. This, combined with the fact that it was also underwater for a long period, has shaped a characteristic topography of rounded mountain slopes.

The first sub-aerial lava flows may go back as far as 20 million years, though the most precise dating indicates an age of 17 million years. The volcanic periods are grouped into four Series or, alternatively, two Volcanic Cycles and what is known as Recent Volcanism.

Material generated by Series I (Ancient Series or First Volcanic Cycle), comprising very old basaltic tongues (Miocene-Tertiary Era) is found throughout the island. This first effusive stage takes in a period from between 17 to 12 million years ago. During this extensive span, fissural basaltic eruptions took place giving rise to large tabular structures formed by the piling up of thousands of lava flows. These first formations are now unmistakable, and are still

Location of the Canary Islands. Dominant relief lines.

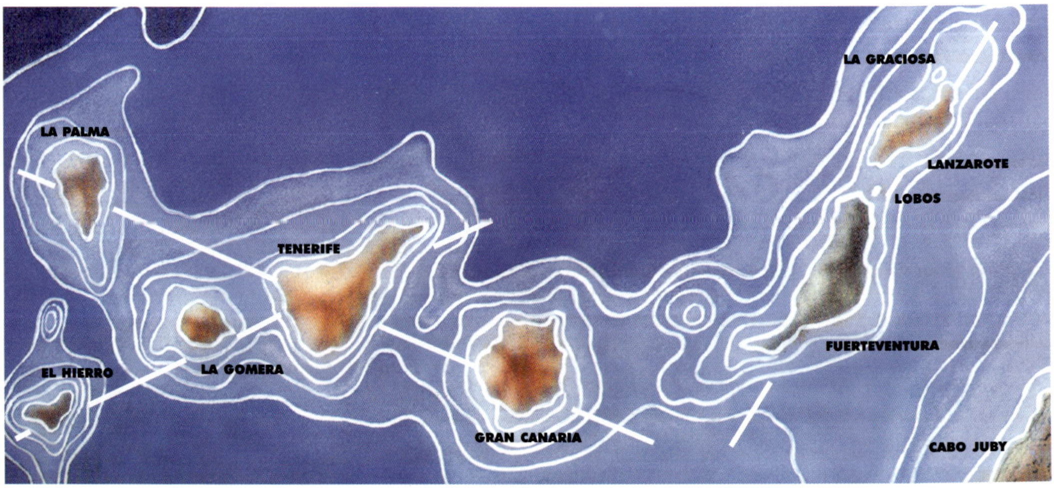

Central-western region of Fuerteventura. Basal complex. Betancuria municipality.

Basaltic rock and solidified sand eroded by the sea on Cofete beach.

Fossilised sand structures.

the highest volcanic structures on the islands despite the intensive erosion to which they have been exposed. The Jandía peninsula is one of these isolated formations formed by the build up of lava, and is characterised by the presence of numerous criss-crossed dikes. What is now the peninsula was once an island and became linked to the rest of Fuerteventura by a more recent and relatively low isthmus.

The build up of lava flows led to the formation of steep escarpments, softened somewhat by ongoing rockfalls and cross-cut by the ravines, which led to the formation of isolated plateaus and jagged peaks. Perhaps the most characteristic feature of the Ancient Series are the U-shaped valleys, some of which were later filled in by other eruptions which changed the nature of the pre-existing topography (Pozo Negro, Barranco del Valle de la Cueva, etc.).

Following a long dormant period, volcanic activity resumed some five million year ago, and the eruptions were to continue almost up to modern times (0.01 million years, Pleistocene - Quaternary Era). This effusive phase spanned various periods (Series II and III, or the Second Volcanic Cycle). Series II affected a wide area in the north and centre of the island and produced formations with a much smoother morphology than its predecessor. The lava spread out widely and formed plains. Associated with this basaltic Series is a thick limestone crust (caliche), occasionally covered with a clayey soil. This layer of caliche - calcium carbonate (CO3Ca), shows that the climate was once more humid than it is today. This Series and Series III, mainly distributed through the northeast part of the island, are associated with very fragmented lava tongues and make for very rocky terrain.

Series IV or "Recent Volcanism", occurring only several thousand years ago, includes the more recent eruptions on Fuerteventura which formed the "malpaises" (volcanic lava). The main such areas are in the north and centre of the island.

The low degree of rejuvenation in Fuerteventura's geological landscape has made the island one of the best existing examples of volcanic morphology reshaped by a long process of erosion. In addition, the scarcity of vegetation allows us to see clearly the successive geological phases which contributed to mould and shape Fuerteventura.

Another noteworthy aspect of Fuerteventura's geography is the abundance of emerged beaches (emerged abrasion platforms), which reflect fluctuations in sea level causing the shorelines to advance and recede. These extensive and shallow abrasion platforms furnished a habitat for numerous organisms, whose remains have, over time, become one of the more important components of the white sands. These sands, accumulated on the beaches, particularly at the present low tide levels, are carried inland by the sea breezes and deposited in dune fields called "jables", the main examples being in Jandía, Corralejo, and El Cotillo.

GEOLOGIC TIME AND FORMATIONS. The geologists contain the history of the Earth in three eons:

Eon	Era	Period	Description	Epoch	Time
EON ARCHEOZOIC Ancient Times			The world already had some continental rock, an ocean, and an atmosphere. The life appears 3.500 m.yr ago in form of bacterias and algae.		4.600-2.500 m. yr.
PROTEROZOIC EON Era of the Primitive Life			The big continents and the extensive seas appeared. 1.800 m.yr. ago, the algae produced the enough oxygen as so they began to accumulate in the sea and in the atmosphere.		2.500-600 m. yr.
PHANEROZOIC EON Period of Geologic time of Visible Life (600 m. yr. until today) Complex life and modern forms. This Eon is divided in successive Eras according to those groups livingin different times.	**Paleozoic or Primary Era** (600-248 m.yr.) The Eras are divided at the same time in periods:	**Cambrian**	All the life was limited to the water. It was, in general, warmer than the current time. The mountains began to grow and the volcanoes threw big plates of lava.		600-505 m.yr.
		Ordovician	Contraction of the oceans and appearance of the first continents. Trilobites lived, primitive corals, and many more invertebrates. Fishes appeared without jaw.		505-440 m.yr.
		Silurian	Marine silts produce rich reservations in petroleum. First plants and terrestrial animals.		440-408 m.yr.
		Devonian	Fossils found in the rocks from this period include river fishes and the first amphibians. The first forests also appear.		408-360 m.yr.
		Carboniferous	It takes the name from the coal strata which produces carbon. they are residuals of tropical forests. Norteamérica came into collision with South America and Europe with Africa. Great part of Antartic, Australia, South America, Africa, and India are covered by ice.		360-286 m.yr.
		Permian	The union of the continents form great continents. The glaciers withdraw toward the south.		286-248 m.yr.
	Mesozoic or Secundary Period (248-65 m.yr.) The Era of the dinosaurs	**Triassic**	Fracture of the great continente. Big amphibians (1 m.), and advanced reptiles (1'5 m.).		248-213 m.yr.
		Jurassic	Warm climate. The primitive birds and mammals appear. Big herbivorous dinosaurs. The current continents have being formed		213-144 m.yr.
		Cretaceous	The deposits from this period include 60% of the reservations of petroleum known nowadays. The plants with flower and the big carnivorous dinosaurs appear. Massive death of dinosaurs and other creatures, perhaps for the impact of a great asteroid.		144-65 m.yr.
	Cenozoic or Terciary Period (65-2 m.yr.)	**Paleogene.** (65-24'6 m.y.)	The continents take their current form. The birds and the mammals occupied the Earth. It is divided at the same time in three Period:	**Paleocene**	65-55 m.yr.
				Eocene	55-38 m.yr.
				Oligocene	38-24'6 m.yr.
		Neogene. (24'6-2 m.a.)	The crashing plates has been elevating the current mountainous chains. The movements of plates, the fault or the volcanic eruptions influenced in the formation of countless islands. **Formation of Canary Islands.**	**Miocene**	24'6-5 m.yr.
				Pliocene	5-2 m.yr.
	Anthropic or Quaternary Era. (2 m.yr.- present time.)		Repeated advances and setbacks of the ice. The last block of ice have melted since 10.000 year ago. The sea levels rise. The man developed. The fauna and the flora diminished.	**Pleistocene or Glacial Epoch**	2-0'01 m.yr.
				Holocene or Recent Epoch	10.000 years Present times.

THE ISLAND'S RELIEF

Fuerteventura is the longest of the Canary Islands, stretching for approximately 100 km from Punta de la Tiñosa, in the north, to Punta de Jandía, in the south. It is the second largest in size, with an total surface area of 1,660 km², including the 4.38 km² of Lobos island.

It is also the closest of the group to Africa, with less than 90 km separating it from the continent, between Punta de la Entallada and Cabo Jubi in Seguía-el-Hamra, Western Sahara.

Off the northeast end of the island lies the tiny island of Lobos, so named for its population of "sea wolves" (lobos de mar), i.e., seals, in the past. This islet is separated from Fuerteventura by a 2 km-wide strait known as El Río, whose depth varies from 9 to 13m.

Fuerteventura is separated from Lanzarote by the 10 km wide Strait of Bocaina with an average depth of 35-40 m.

The distance between Punta de Jandía and Punta de Gando, in Gran Canaria, is about 80 km.

Solidified sand formations.

Fuerteventura is located in the Northern Hemisphere between 28° 45' 04" and 28° 02' 1" Northern latitude, and between 13° 49' 12" and 14° 30' 24" Western longitude.

The present landscape is the outcome of the blend of two opposing forces, the constructive power of volcanic eruption and the destructive force of erosion.

As mentioned above, the island has a markedly mature topography with few mountainous formations. The maximum height is Pico de la Zarza at 807 m, on the Jandía

Accumulated sand giving rise to large dunes in the southwest of the island.

Barranco de la Peña.

Peninsula. Betancuria Massif (Basal Complex), is the most mountainous relief feature, and reaches its maximum height of 762 m at Pico de la Atalaya.

In the newer northern area, there lies a lava field ("malpaís") dating from Quaternary Era volcanoes. The most outstanding of these volcanoes are Montaña la Mancha (151 m), Montaña San Rafael (117 m), and Montaña Bayuyo (269 m). Also worthy of mention is the lava field known as Malpaís de la Arena, between Lajares and La Oliva, formed from sand and lava from Montaña de la Arena. Inland, in the area of Tiscamanita and Tuineje, recent eruptions from Caldera de Gairía (461 m) led to the formation of Malpaís Chico; while material erupting from Caldera de la Laguna (301 m), Caldera de Liria (254 m), Caldera de los Arrabales (242 m), and La Caldereta formed the Malpaís Grande.

Roque del Moro. Cofete.

Also worth mentioning in the northeast are Montaña Roja, a solitary elevation of 312 m, and Montaña Escanfraga, Fuerteventura's highest volcano at 531 m.

Large areas are covered by organic sand formations and dune fields, known as jables, the main examples being in Corralejo, northeast of La Oliva and at Istmo de la Pared (La Pared isthmus), in the south. The wind effect has played an important role in shaping the island's coasts and has been aided by the large number of expansive open beaches.

The Betancuria massif is located in the west of the island, along with the three most important central ravines: La Peña, Ajuí and Pájara. Barranco de las Peñitas flows into the

Cliffs in the northwest.

latter and is a deep canyon with a place in history, since Béthencourt and Gadifer de La Salle passed through it on their way inland from Puerto de la Peña during their conquest of the island. No less historic is the "hermita" (chapel) de Nuestra Señora de la Peña, patron saint of Fuerteventura; and the Las Peñitas reservoir, in the Mal Paso defile.

From Montaña Roja to Pozo Negro in the east, there is a succession of low terraces with numerous beaches, such as Playa Blanca, El Matorral, and Caleta de Fuste.

From Pozo Negro to Istmo de la Pared, the shore is cliff-lined, alternating with coves and beaches at the mouths of the ravines; Las Playitas, Gran Tarajal, Giniginámar, Tarajalejo, all the way to Matas Blancas where the extraordinarily beautiful beaches of Sotavento (leeward coast) Jandía begin, with their dunes of white sand and a multicoloured sea.

Also in the east, Cuchillo de la Entallada, at 283 m, looms over the sea and is the island's closest point to Africa. There is also a lighthouse, which is noteworthy for its unique design.

Valle de Río Palmas. Basal complex.

Die Ebene Llanos de Casillas del Angel.

The great plains inland from the coast are among the most characteristic of the island's landscapes, surrounded by the mountainous relief of Betancuria and the eastern ridges. They are the so-called "vast landscapes", with a markedly fragile ecosystem due to aridity and wind erosion.

In the southwest, the peninsula of Jandía constitutes a separate unit from the rest of the island. It begins with the Istmo de la Pared, a low isthmus covered with sand and dunes. In this area, we find the largest and most spectacular beaches not only in Fuerteventura but in the entire Archipelago. In the north of the peninsula lie the Barlovento de Jandía beaches, followed by the beach of Cofete as far as Roque del Moro, after which the coastline becomes abrupt until Punta Pesebre, next to the cliff-lined Caleta de la Madera. This area is one of the biggest yet little known places of Fuerteventura. Between Punta Pesebre and Punta de Jandía and its lighthouse,

Malpaís Grande and the calderas which formed it: La Laguna, Liría and Los Arrabales.

lies the Ojos beach. From Punta de Jandía until the town of Morro Jable, the coastline is varied, with alternating low cliffs and beaches at the mouths of the ravines. From Morro Jable to Punta de los Molinillos at the isthmus, the coast is made up of the Sotavento de Jandía beaches.

Numerous, large ravines cut across the peninsula, originating on the mountainous spine that separates the windward and leeward coasts, and flowing towards the Sotavento or leeward shores; among them, Ciervo, Canarios, and Mal Nombre. Also of note are the Esquinzo and Butihondo ravines emanating from the Pico de la Zarza.

As already mentioned, the coastline is highly varied - there are low terraces and cliffs alternating with innumerable beaches of the most diverse textures and colours.

The coasts are divided into Barlovento and Sotavento, in accordance with the direction

Dune fields at Corralejo

of the prevailing winds. Barlovento means against the wind - the wind is coming from the sea; Sotavento means the wind is at your back - or offshore. The Barlovento seas are choppy, whereas the Sotavento ones are noticeably placid, making them a haven for swimmers and watersport enthusiasts. The main tourist centres are also found along the leeward coast.

In terms of size, access and facilities, the following are among the best beaches:

El Cotillo.

Playa Blanca: A beach of golden sand, located between the airport and Puerto del Rosario. Behind the beach there is a string of dunes anchored by vegetation - "balancones" which continue inland as far as the slopes of the nearby ravine of Río Cabras.

Caleta de Fuste: About 10 km south of the airport. A golden sandy beach with a natural inlet open to the southeast. The tourist centre of El Castillo, one of the island's most important, is based around this beach. All facilities and services are available. There is a major recreational port, and diving and windsurfing centres.

Playa de Gran Tarajal: A black sand beach situated next to the town of Gran Tarajal at the mouth of the ravine of the same name. Behind the beach there is a small ravine with a brush typical of marshes. Inland, the valley is home to palm and tamarisk trees, arranged around the perimeters of old "gavias".

Playa Blanca.

Playa de Gingināmar: A beach of black sand and pebbles next to the fishing centre of Giniginámar, at the mouth of the ravine of the same name. The beach is backed by some palm and tamarisk trees and an open brush of "matomoros" (Suaeda vera).

Playa de Tarajalejo: A black sand beach, larger than the beaches of Gran Tarajal and Giniginámar, next to the town of Tarajalejo at the mouth of the ravine of the same name. The valley has a palm grove similar to that of Gran Tarajal, though somewhat smaller. The proximity of the old fishing village and the surrounding

Caleta de Fustes.

Gran Tarajal - beach and town.

tourist resorts ensure the availability of all facilities and services.

Playas de Sotavento de Jandía: At the base of the southeast slope of the Jandía massif lie the longest beaches of the Canary Islands. This golden sandy coast includes the beaches of Matorral, Butihondo, Esquinzo, Mal Nombre and Sotavento, stretching from Costa Calma to Morro Jable. The island's main tourist resorts are found along the length of this coast. The endless possibilities for leisure and tourism are complemented by an outstanding natural setting. Large dunes, pools, ponds, and salt marshes, and the life forms associated with them, constitute a complex interrelated whole. The shoreline advances with the creation of sandbars running parallel to the coast, leaving behind depressions that are periodically flooded, out of which rise mounds crowned with "uva de mar" and "balancones". Further in and connected to these ponds lies an area of shallow waters with a

Giniginámar - beach and town.

Tarajalejo - beach and town.

complex network of channels. The area is rich in flowers and is frequented by birds. These salt marshes are of great ecological interest, the scrubmarsh here being considered as the most important in the entire Archipelago

Playa de Juan Gómez: A small beach of golden sand lying at the foot of the escarpment at the mouth of the Juan Gómez ravine. It can be reached by taking a turnoff 300 m before the junction in the road linking Morro Jable with Jandía Point and Cofete. This 400 m long beach is also favoured by the sheltered leeward conditions. The privileged natural surroundings

Sotavento de Jandía.

Playa del Matorral.

also include other interesting coves. At the top of the escarpment there are some isolated patches of "uva de mar", and further inland, a thorny brush grows on the mixed caliche and sedimentary slopes.

Playa de Los Ojos: A small beach of golden sand at the base of an escarpment, with transverse dykes jutting into the sea. Further inland lies the Llanos del Cotillo, or La Angostura, a vast stretch of solidified sand with small dunes anchored by plants. Access to the beach is by a road which runs from Puerto de la Cruz (near the lighthouse) to Punta Pesebre in the extreme northwest of the island. The cliff is about 10 m high, and you must be very careful on your way down.

Playas de Barlovento de Jandía. Cofete: Between Roque del Moro and Punta Paloma, at the base of the northwest slope of the Jandía massif, there lies a 14 km stretch of golden sandy beach. As mentioned earlier, the windward

Playa de Sotavento de Jandía.

Sotavento de Jandía. Barranco del Salmo.

shores are rough and the breeze strong, which means that great caution must be taken with watersports. To reach this coast, leave Morro Jable by the fishing port exit. After 40-50 minutes, the houses of Cofete - where you can have lunch - come into view. The deserted beach is ideal for those in search of tranquillity and open space. The beaches of Cofete and Barlovento (separated by the Islet) lie in one of the most impressive geological settings of Fuerteventura - at the foot of huge cliffs crowned by the Pico de la Zarza (807 m), the highest point on the island. Beyond the "balancones" on dunes anchored by thorny brush, on the steep slopes of the massif, you can just make out some patches of thistle. Further up, on the heights favoured by the constant humid winds, there thrives an authentic botanic treasure.

Sotavento de Jandía.

Playa del Viejo Rey: This beach marks the northern extreme of the area where the sands are blown in, giving rise to the dune fields of Istmo de la Pared. A beach of golden sand at the foot of the escarpment, there are large wind-blown dunes rising towards the interior. To the south lies a string of small beaches of similar characteristics, in some cases visible only at low tide.

Playa de la Pared: Located next to the beach of Viejo Rey, this beach is a marked contrast to it. It is a beach of black sand at the mouth of a ravine carved into the slopes of the Betancuria massif. This beach includes a number of tourist facilities.

Playa de Juan Gómez.

Los Ojos.

Playa de El Aljibe de La Cueva: Jarugo, Tebeto, Esquinzo, El Águila and El Aljibe de La Cueva form a set of beaches along the coast of the Betancuria massif. The beach of El Aljibe de La Cueva, near El Cotillo, is the largest of them at some 2 km long and is similar to the others. These beaches are set at the foot of a 20 m high cliff at the mouths of ravines which flow all year round, which makes them ideal habitats for birds. The elevated beaches and emerged marine shelf, which stretch from the edge of the cliff to the base of the massif, bear witness to the changes in sea level. The northern sea fosters a great biological richness seen primarily in the abundance of fish and mussel colonies. The same winds and waves that protected and nurtured this treasure are nowadays an added attraction for all kinds of watersports.

Playas de El Cotillo: Along the north coast of El Cotillo, around the El Tostón lighthouse, there are a number of small coves with

Barlovento de Jandía. Cofete.

Viejo Rey.

beaches of shiny white sand: Caleta del Río, Caleta de La Aduana and Caleta del Marrajo, among others. These coves can be reached via the road leading to the lighthouse. The winds are usually constant and the waves break a good distance from the shore, leaving vast areas of calm, crystal clear water. Along the shores, mud-dwelling birds are often seen resting or feeding on the seaweed washed in by the tides. The islets of sand are covered with a vegetation which is very different to that of the malpaís, with species such as "balancón", "lecheruela", and "salado", etc. The luminosity and colour contrast arising from the meeting of lava, sand and sea give rise to a panorama of striking, ideal beauty.

Playas de Corralejo: A group of yellow sandy beaches bordering the Corralejo Nature Park, between Montaña Roja and Corralejo. From the coast, a dune field stretches inland, burying part of the La Caldera Encantada malpaís. The road that runs from Puerto del Rosario to Corralejo takes us close to these beaches, offering stupendous views along the way. Some huge dunes have spread to the other side of the road, while others have remained closer to the shore, anchored by "balancones". As with the beaches of El Cotillo, a magnificent range of colours dominated by emerald tones adorns the shores, which are calm for most of the year. These beaches, together with those of Jandía, are the

La Pared.

Cofete.

most spectacular of the island and for that reason tend to be the most frequented by tourists. They boast a complete range of services and facilities, with steady winds ideal not only for windsurfing but also for kite flying - a popular and colourful pastime. Divers will also delight in the spectacular sea beds of the strait between the beaches and Lobos island.

Another outstanding geomorphologic feature of Fuerteventura is its underwater world. The island is blessed with the most extensive platform in the Archipelago. This platform is connected to the island of Lanzarote to the north, and stretches south several kilometres beyond Punta de Jandía. On both sides of the central part of the island, in front of the La Entallada lighthouse and Punta de la Herradura, it takes the

El Aljibe de la Cueva.

form of extensive shoals, known as "placeres" by local fishermen.

The seabed of the platform is mostly sandy to leeward, becoming rocky closer to shore and along the edge, which the sailors call the "veril". Large sandy areas covered with green plants of the genus Cymodocea prevail up to a depth of 30 m These underwater meadows are called "manchones" or "sebadales". In the rocky areas, there are vast stretches of compressed sand bottom covered with flat rocks which the fishermen call "tableros". There are also smaller

La Pared.

areas of coral bed; these are always found close to the veril, at depths of about 150 m.

Along the "Barlovento" or windward coast, the platform is much rockier and green areas are few and far between; the rocks on the sea bed are very tall, in some cases up to 20 m. These are known as "bajones" and the tops are visible on days when the water is clear. This type of rocky seabed with its highly accentuated relief and numerous caves, is known as "marisco

Playa Negra.

Playas de Corralejo.

bronco" in Fuerteventura. Relatively even and low lying rocky beds are known as "marisco atablerado".

The offshore platforms slope gently to a depth of about 200 m, after which they break off and plunge dramatically to the great depths. In some cases the drop is almost vertical, as off La Lajita. At the foot of the banks, at a depth ranging from 500 to 600 m, there begins an relatively flat area, with a muddy bed and isolated rocks.

All this marine world is washed by the ocean currents. which are particularly persistent and intense in the Sotavento sea, flowing from northeast to southwest, parallel to the island.

Playas de El Cotillo.

Playa del Aguila.

NATURAL RESOURCES

Climate. The climate is semi-desertic, characterised by low rainfall (140 mm per year) and the constant presence of the trade winds. However, the climate is moderated by the sea, with its cold Canary current, and average annual temperature is 20.16º C, with relatively little difference between minimums and maximums. The coldest month is January and the hottest, August.

Due to its subtropical location, the Canary Archipelago is open to anticyclones (high pressure air belts) for the greater part of the year and especially in summer. The anticyclone affecting the Canaries is the Azores anticyclone, and it gives rise to the trade winds. These become laden with humidity and salt on their way to the

islands, and frequently bring horizontal cloud patterns. Their speed increases on meeting the obstacles presented by the islands' relief, reaching an average of 23 km per hour.

When the anticyclone disappears, the winds change direction and origin. They may originate in the Atlantic (SE) or in the Polar region (N or NE). Polar winds give rise to the minimum annual temperatures, what is known

El Cotillo.

as the "gota fría" (cold air) effect, and the bulk of annual rainfall; alternatively, the winds may have tropical continental origins (i.e., Africa), in which case they come laden with suspended dust particles in what is known as "tiempo sur" or "tiempo levante" (literally, southern or eastern weather).

Relative humidity oscillates between 19% and 38%. And with an average of 2,800 hours of sunshine per year, Fuerteventura is an outstandingly sunny place. Rainfall is scarce and mainly occurs from September to May. Storms are rare. Sea water surface temperature is between 17 and 23 º C.

Settlement and Development.

The Canary Islands are part of the bio-geographical area known as Macaronesia, a term derived from the Greek (Makaros = happy, and nesos = island). Macaronesia comprises five Atlantic archipelagos (Madeira, Azores, Salvagens, Canary Islands, and Cape Verde). The area, taken as a whole, is marked by characteristically similar flora and fauna.

Mists rising over Cofete.

Relative humidity is the percentual quotient among the quantities of water vapor from the air to a certain temperature. And the maximum quantity of water vapor the air could contain to that temperature.

The **compass-card** consists on placing the corresponding frequency -expressed in % - for each direction of the wind to every month, obtaining the monthly average. Or for every year, being the result of the annual average compass-card. In the center the percentages of calm are placed using the same scale. For Fuerteventura, the anual compass-card is:

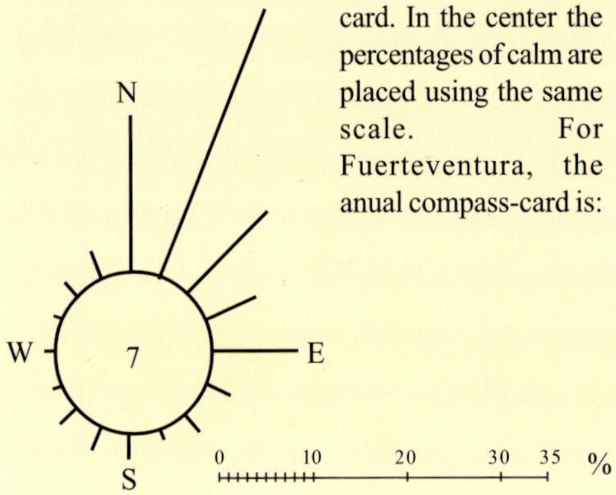

The majority of the Archipelago's plants and animals are of African and European origin. While the continental flora and fauna adapted to profound climatic changes over the last few million years, the Atlantic archipelagos of Macaronesia maintained a great diversity of habitats and a relative climatic stability. These conditions permitted the survival of species that were disappearing from surrounding areas, and in some cases, from the rest of the world; at the same time, new species evolved from the ones that had arrived earlier.

The Canary Islands became a living museum, with species and ecosystems of an earlier era, and abundant examples of island evolution and the creative capacity of nature. Had it not been for the famous quarantine that prevented Darwin from reaching the islands, he would have found sufficient arguments for the mechanisms of evolution well in advance of the discoveries he would later make in South America and the Galapagos Islands. While vertebrate fauna is not abundant on the islands, there are nevertheless, among the reptiles and birds, similar examples to those used to demonstrate his famous Theory of Evolution. In contrast to the relative poverty of vertebrate fauna, there is a veritable treasury of plant and associated invertebrate life, so that even today new species are being discovered and added to the world's natural patrimony.

The tremendous interest generated by nature in the Canary Islands has been recorded and documented by prestigious researchers over the years (Humboldt, Webb, Berthelot, Viera y Clavijo, etc.). In addition, many of the islands' natural habitats have been declared Human Heritage Sites (El Cedro), or World Biosphere Reserves (Los Tilos Forest and the island of Lanzarote).

Flora. Although Fuerteventura's endemic flora is not abundant in comparison with the other Canary Islands, which account for 650 endemic species in a surface area of 7,200 km^2, it is nevertheless extremely rich when compared with any European country. France, for example, with a surface area of 560,000 km^2, has some

Cardón Canario (Euphorbia canariense)

Cardón de Jandía (Euphorbiaceae, Euphorbia handiensi).

100 endemic species, the U.K., with a surface area of 250,000 km², accounts for 16, while Germany with 350,000 km² has only 6 endemic species. In this light, the wealth of Fuerteventura's flora takes on another dimension.

Jorjado o tojia (Odontospermun sericeum).

The island's vegetation pattern takes the form of a number of different levels, the combined result of the constant flow of the humid trade winds and the height of the land masses. Even at their highest, Lanzarote and Fuerteventura do not exceed 700 m above sea level, and as a result the winds usually pass over without releasing their moisture. In contrast to the forests found on the higher islands of the Archipelago, Fuerteventura offers the best examples of semi-desert habitats of the Canary Basal Complex. The island's vegetative covering is rather poor, due not only to the prevailing aridity, but also to over-grazing; there are also volcanic zones where the natural plant colonisation process has not yet had time to progress beyond the early stages.

Nevertheless, in the past, Fuerteventura was the grainyard of the Canary Islands as well as having important goat raising activities (which still continue). The decline of agriculture in general has allowed gradual recovery of plant

Lichen on the malpaís.

Tabaiba salvaje (Euphorbia obtusifolia).

species which were on the verge of extinction. Today's landscape is a dynamic one, slowly recovering its original form of vegetation which until recently was only barely existent in inaccessible or unproductive areas.

Thorny scrub is the plant form occupying most surface area on the island, growing on the plains and low hills, as well as in untouched and previously cultivated areas. In this recolonisation process, the hardy furze ("aulaga" - Launaea arborescens) has become one of the main elements of the advancing scrub and is now one of the most common species on the island, to the point of dominating certain areas. Some of the better preserved areas exhibit a greater diversity of plant life, including hawthorn ("espinos" - Lycium intrincatum), "matomoros" (Suaeda vera), and "rama" (Rhamnus crenulata), among others.

"Verode" (Senecio kleinia), "tabaiba" (Euphorbia balsamifera) and "cardón" (thistle - Euphorbia canariensis and handiensis) are the

Lichens: «Ramalina» (green) and Xanthoria (orange).

purest representatives of the original vegetation, and are believed to be indigenous. The tabaibas generally grow on hills and escarpments, though some have taken to the more recently formed lava fields. The "tabaiba dulce" ("sweet tabaiba") tends to establish itself in areas open to the ocean breezes.

The "cardonales" (thistle fields) tend to be restricted to the Jandía area. In addition to the species which is common to the entire Canary Archipelago, there is also a local endemic variant only found in a small area on the peninsula. Beautiful and of scientific interest, they are also highly fragile, due to their scarcity locally and

Gatuña, melojo o cueso (Ononis natrix).

Canary palm (Palmae, Phoenix canariensis).

worldwide. In the case of the "Cardón de Jandia" (Euphorbia handiensis), this scarcity is extreme indeed. The "Cardón canario" (Euphorbia canariensis) also accommodates other plant forms in its interior, e.g., "cornical" (Periploca laevigata), "tasaigo" (Rubiafruticosa), "espina blanca" (Asparagus pastorianus), thus protecting them from the foraging goats. The survival of a great number of invertebrate endemic species is directly dependent on this candelabra-shaped cactus-like member of the Euphorbiacea family.

The palm and tamarisk trees are practically the only examples of indigenous tree forms. They are usually found in ravines, valleys and especially, along the old cultivated terraces known as "gavias" which optimise the use of land and water.

In the ravines with running water, thick brush grows inundated with "matamoros". This plant and others are members of a richer community: the "saladar" or salt marsh. In addition to their botanical importance, represented by species such as "salado" (Schicozogyne glaberrima), "saladillo" (Jubera canariensis), "mato" (Lavandula canariensis) and "tebete" (Beta patellaris), the importance of the salt marsh lies in its role as a wetland, and thus, as a setting for bird life.

Close to the salt marshes, we also find dune fields. The dunes stretch inland, blown by the prevailing winds. The tremendous aridity and friction caused by the movement of sand, allied to the effects of sun and salt, greatly limit the possibilities for living beings. Among the species that do take hold under such conditions are the

Uva de mar (Zygophyllaceae, Zygophyllum fontanesii).

Canary basal floor. Semi-arid mediterranean climate vegetation with rains among 100-300 mm. 20 º C average temperature. High insolation. With loamy, saline or calcareous floors (*caliches*). Example: arid pastures, *heaths xerófilos* (storing water plant), euphorbias canariences, euphorbias sp., arthrpchemum fruticosum.

"uva de mar" (literally "sea grape" - Zygophyllum fontanesii) with its fleshy Y-shaped leaves, and the "balancones" (Trajanum maquinii), sometimes extremely large and invariably found on established dunes behind the beach. Inland, the sand supports a variety of plant forms including "corazoncillos" (Lotus lanzarotensis), "algahuera" (Cheroloides tormentosa), "salado blanco", "saladillo", "criadilla" (Erifezia pinoyi), "melosa" (Ononix natrix), and "cebollín de playa" (literally "little beach onion" - Hyrocymbium psammophilum), among others.

The undoubted wealth of Fuerteventura's plant life is especially characterised by its low visibility in the island's landscape. A few treasures are found on the highest points of Jandía, with endemic species and vestigial "monteverde" vegetation ("Joraos" - Odontospermium seniceum, "magarza" - Chysanthemum, "tajinaste de Jandía" - Echium handiensis, "acebuche" - Olea cerasiformis); while others go practically unnoticed, such as the lichen colonies comprising a wide variety of species, which form multi-coloured crusts on the rocks. Two lichens with a somewhat more complex structure are "aicán" (Roccella) and orchilla, formerly exploited as a source of dye. These lichen colonies are the main features of the wide rocky fields stretching over the plains, hillocks and slopes, the rocky escarpments, and of course, the "malpaises" or lava fields.

Fauna. The invertebrates, and especially insects, are the best represented fauna in the Canary Islands and in Fuerteventura. More than 5,000 species have been catalogued

View of the north from Morro Velosa.

Antophoras.

Cochinillas.

in the Canary Archipelago, with close to 10% of these being endemic. In the case of the beetles, the number of endemic species is as high as 35%. Of the insects found in Fuerteventura, two are particularly noteworthy and play a role in the island's landscape: one is a fossil, the other is living.

The antophoras bee was similar to the present-day wasp. They tunnelled out a cavity in the sand for their larvae. The fossilised remains of these nests are found today in great quantities in the inland sands, where they appear as small clumps of compressed sand, hollow and ellipsoidal.

The small white spots that appear on the leaves of the "tunera" (prickly pear-cactus) are the female of the "cochinilla", an insect of formerly great economic importance, for the highly valued carmine dye obtained from it. This insect became protected by law in 1827. The value of such products stands to rise again in the light of studies highlighting the health risks involved in the use of much cheaper artificial substances.

In contrast to the high number of invertebrates, the Canary Islands are not abundant in vertebrates, birds being the only group represented by a considerable number of species. Of the islands, Fuerteventura has the greatest number of birds, with many nesting, migratory and endemic species. Due to the Archipelago's strategic location, as an area of passage and winter stopover at the extreme end of a migratory route, the importance of the Canary wetlands goes far beyond the local level. The virgin coasts of

Great grey shrike

Pipits.

Fuerteventura, with their salt marshes and ponds, provide a welcome stopover for these species, and the components of the community changes with the season.

The most characteristic birds found on the sandy and earthy plains are the houbara bustard or great bustard, the cream-coloured courser, the stone curlew, the great grey shrike, the kestrel, and the hoopoe, among others.
In the ravines, we may find the lesser short-toed lark, the stonechat, the blue tit, the Spanish sparrow, among others. The ravines are also fine habitats for migratory and wintering birds.

Raven

In the mountains there are a variety of nesting birds such as the Egyptian vulture, the buzzard, the trumpeter bullfinch, the osprey (on the coasts and cliffs), the barn owl and others. Both the Egyptian vulture and the osprey are in imminent danger of extinction.

Of the marine birds, the most notable are Cory's shearwater, a nesting bird; the ringed plover, the grey plover, the dunlin, and the little egret, all migratory. Deserving of special mention are the few remaining common terns, a species previously present in great numbers.

On the island of Lobos, nesting birds include Bulwer's petrel, the Manx shearwater, Cory's shearwater and the storm petrel. Some others are found on the nearby rocks and islets.

Barbary partridge

A colony of terns.

The special attention given to the marine fauna for so long is due to a very practical consideration: fishing. More recently, however, there has been a growth in scientific and technical interest. Some 390 species of fish divided into 117 families have been classified around the island. Naturally, it is not possible to mention all of them here, but we can at least give a few facts. Firstly, with regard to the ecosystems and waters surrounding Fuerteventura, and secondly, the most representative species inhabiting them.

Due to the ocean currents in the Canary Islands, the local waters present anomalies of salinity and temperature which greatly influence marine life. These waters permit the existence of fish from diverse faunal regions. We can identify two distinct ecosystems, the "Barlovento" or north sea which, conventionally, is situated between El Cotillo and Punta de Jandía, and the south or "Sotavento" sea, which lies between Punta de Jandía and Corralejo. Between Corralejo and El Cotillo there is a completely open stretch of water that can be considered a transition area.

Kentish plover.

Grouper

Moray surrounded by cardinal fish.

The differences between the two zones do not lie solely in the nature of the seabed, sandy as opposed to rocky, ("marisco" in the fishermen's jargon), or in the surge and swell that affect the windward side for most of the year, but rather in the quality of the waters. On the north side, the water is slightly cooler and much richer in plankton. This results in a greater abundance and diversity of marine life and also a greater variety of species. Also the north sea is less fished and has a number of characteristic fish species.

The majority of the fish species are "Creole", spending their entire life cycle in the island's waters. The remainder, "travelling fish" as the seamen say, are frequent visitors: tuna, gilthead bream, sardines, snipefish, and various species of sharks, which are found there at certain times of the year.

By families, the sea breams predominate: bogues, salemas, white sea bream, black sea bream, saddled bream, red banded sea bream, pink dentex, common pandoras, common sea bream, axillary sea bream, among others. These species are the typical species of the platform.

The next most represented species is the shark. The majority are "Creole" and live close to the seabed. Samples are the smooth-hounds, spurdogs, gulper sharks, kitefin sharks, and blackmouth catsharks, among others. The most abundant of the "travelling" sharks are the scalloped hammerheads, shortfin makos and blue sharks.

In the "mariscos" or areas with rocky seabeds, the groupers are particularly well represented, not so much for variety as for abundance: dusky groupers, combers, comb

The sandy seabeds give a turquoise colour to the coastal waters.

Mussels (Perna perna).

groupers; and the wrasses: ornate wrasses, rainbow wrasses, scale-rayed wrasses, and others.

Tuna and similar are represented mainly by skipjacks, Atlantic bluefins, yellowfins and Atlantic bonitos, in addition to yellowmouth barracudas and the highly abundant chub mackerel.

Of the species not included in the above-mentioned families, the most important are:

On the platform: lesser weevers, African striped grunts, alfonsinos, grey triggerfish, thinlip grey mullets, fangtooth morays, striped red mullets, fusca drums and parrot-fishes.

On and around the verges: conger eels, stone basses, silver scabbardfish, mora moros, and deep-water alfonsinos.

The pelagic species include halfbreaks, flying fish, dolphins, blue fishes, swordfish, sardines, and snipe fish.

The crustaceans are represented by a multitude of species. The best known are:

Male lizard.

Among the crabs: the "santorras" (spider crabs) which are frequently caught in shallow-set traps or "nasas"; the white and red crabs, widely sought after for their exquisite meat, caught by hand from the shore on moonless nights at low tide. The "juyón" or "carnada de vieja" (Xantho porssa), found under rocks on the shore and used as bait to catch parrot fish and red banded sea breams.

Shrimp are found in great abundance.

The Canary "bogavante" or lobster is very common in sea caves on the northern side. Barnacles proliferate in the shady areas of the northern sea reefs. The older people call them "patas de cabra" (goat feet).

The most abundant shellfish is the mussel and the Barlovento coast of the island is covered with them. Sea snails ("burgaos" and "cañaillas") and limpets also abound.

The colourful "orejas de mar" ("sea ears") can be found under rocks near the shore. The islanders refer to them as "almejas" or clams, though they have only one valve. Endless varieties of other sea snails are also to be found.

Among the cephalopods, octopus, squid and cuttlefish are especially abundant.

Further offshore, there are many sea turtles, especially the "boba" turtle (loggerhead - Careta careta), and occasionally laud turtles, which nest on some of the island's beaches.

The most common marine mammals are the "toninas" (dolphins), some whales, and in recent times, the occasional seal.

The amphibians are represented by the common frog, which requires moist freshwater areas such as ponds and ditches to survive.

Barbary ground squirrel.

Goats

Reptiles are present in the form of lizards, the "lisnejas" or "lisa majorera" (thought to be endemic to Lanzarote and Fuerteventura), and geckos.

All the island's mammals were introduced, voluntarily or involuntarily, by man, with the exception of the bats, who managed to get there on their own. This includes domestic animals like dogs (the bardino being an autochthonous breed), goats, pigs, sheep, horses, dromedaries, rabbits, donkeys and cats.

Non-domestic mammals are found throughout the island. These include hedgehogs, shrews, bats, rabbits, mice, rats and barbary ground squirrels, the latter a recent introduction from Africa currently inflicting significant damage on the island's agriculture.

Many specialised studies have been carried out and reported on the invertebrates, reptiles, birds and fish of Fuerteventura.

Protected Natural Areas. The Canary Islands are a "Macaronesic" archipelago in which the climatic, geological, marine, geomorphologic, zoological and botanical characteristics have constituted an exceptional natural environment. However, the regional fragmentation of the islands, their demographic density and the almost exclusive economic dependence on the service sector have generated a form of development which is unsustainable. The capacity of the natural resources to recover is simply being outstripped by the pace of exploitation.

Remedying this situation will require a special effort in education and environmental consciousness-raising.

The "Ley de Espacios Naturales de Canarias" (the Law on Canary Island Nature Reserves) governs the protection of the Archipelago's natural resources and sets out the nature of the planning instruments.

The Law aims to ensure a reasonable use of natural resources leading to sustainable development; the integration of areas in need of preservation and restoration; promotion of scientific research; environmental education and sustainable human use of nature; the improvement of the quality of life of the local communities, and restoration and recovery of the islands' ecosystems.

In accordance with the natural resources they contain, the nature reserves are incorporated into a network of the most important natural habitats and centres of biodiversity. The categories are as follows:

- Nature and Rural Parks.
- Nature Reserves: Integral and Special Nature Reserves.
- Natural Monuments.

Jandia Nature Park. Cofete beach.

Betancuria Rural Park.

- Protected Landscapes.
- Places of Scientific Interest.

Nature Parks are large areas, largely unaffected by human exploitation or occupation, whose natural beauty, flora, fauna and geology are considered to be of exceptional value.

Rural Parks are large areas in which agricultural, livestock or fishing activities coexist with features of great natural and/or ecological value.

The aim of the **Nature Reserves** is to protect ecosystems, communities and biological or geological features which, due to their rarity or other characteristics, are especially deserving of respect. The removal of biological or geological materials from these areas is prohibited.

Integral Nature Reserves are smaller areas in which the aim is to protect all their features and natural ecological processes.

Special Nature Reserves are also smaller areas in which the aim is to preserve exceptional habitats, certain species and natural ecological processes of special interest.

Betancuria Rural Park. La Peña reservoir.

Natural Monuments are natural spaces or features, usually limited in scale, comprising natural formations which are either unique, rare or of great beauty. These tend to be geological or paleontological features which are of scientific, cultural or aesthetic interest.

Protected Landscapes are areas of special aesthetic and cultural value.

Places of Scientific Interest are generally relatively small isolated areas containing natural elements of scientific interest, animal or plant species in danger of extinction and in need of special protective measures.

The Protected Natural Areas of Fuerteventura are as follows:

- Nature Park of Islote de Lobos. Comprised of 467.9 hectares in the municipality of La Oliva, it includes the island surface area as far down as the low-tide mark.

- Corralejo Nature Park. Comprising 2,668.7 hectares also in the La Oliva municipality.

- Jandía Nature Park. Some 14,318.5 hectares in the Pájara municipality.

Ajuí Natural Monument. The beach.

- Betancuria Rural Park. Made up of 16,544.7 hectares spread over the municipalities of Betancuria, Puerto del Rosario, Antigua, Tuineje and Pájara.

- Malpaís de la Arena Natural Monument. It comprises 870.8 hectares in the municipality of La Oliva.

- Tindaya Mountain Natural Monument. 186.7 hectares in the municipal boundary of La Oliva.

- Caldera de Gairía Natural Monument..

Montaña de la Arena volcano and malpaís (Recent Volcanism). La Oliva.

Jandía Nature Park. Sotavento beaches.

Comprising 2,400.9 hectares in the municipal areas of Antigua and Tuineje.

- **Cuchillos de Vigán Natural Monument.** Some 6,090 hectares in the municipal areas of Antigua and Tuineje.

- **Cardón Mountain Natural Monument.** Some 1,266.8 hectares in the municipal boundary of Pájara.

- **Ajuí Natural Monument.** An area of 31.8 hectares in the Pájara municipality.

- **Malpaís Grande Protected Landscape.** 3,245.3 hectares in the Antigua and Tuineje municipalities.

- **Vallebrón Protected Landscape**. 1,679.6 hectares in La Oliva and Puerto del Rosario.

- **Playa del Matorral - Place of Scientific Interest.** An area of 115.6 hectares in Pájara.

Vallebrón Protected Landscape.

A visit to these protected areas allows you to see the different types of landscape which are characteristic of the island: sweeping plains, rocky massifs, lava fields, ravines and, of course, the magnificent coasts with their towering cliffs and immense beaches.

It is important to bear in mind the delicacy and fragility of Fuerteventura's ecosystems when visiting these areas. There is an urgent need not only to contribute to improved general information on these ecosystems, but also to channel and control leisure uses so as to protect them. The degradation currently taking place, for example through the proliferation of numerous new pathways, must be eliminated.

If you wish to explore these areas in more depth, we recommend you take one of the routes described later in the book. Almost all of these routes take you through protected areas and this is a great way to enjoy the tranquil landscapes of Fuerteventura.

Jandía Nature Park. La Barca.

Islote de Lobos Nature Park.

1 PARQUE NATURAL DEL ISLOTE DE LOBOS
NATURPARK DER INSEL LOBOS
NATURE PARK OF ISLOTE DE LOBOS

2 PARQUE NATURAL DE CORRALEJO
NATURPARK VON CORRALEJO
CORRALEJO NATURE PARK

3 PARQUE NATURAL DE JANDIA
NATURPARK VON JANDIA
JANDIA NATURE PARK

4 PARQUE RURAL DE BETANCURIA
KULTURLANDSCHAFT VON BETANCURIA
BETANCURIA RURAL PARK

5 MONUMENTO NATURAL DEL MALPAIS DE LA ARENA
NATURDENKMAL MALPAIS DE LA ARENA
MALPAIS DE LA ARENA NATURAL MONUMENT

6 MONUMENTO NATURAL DE LA MONTAÑA DE TINDAYA
NATURDENKMAL MONTAÑA DE TINDAYA
MONTAÑA DE TINDAYA NATURAL MONUMENT

7 MONUMENTO NATURAL DE LA CALDERA DE GAIRIA
NATURDENKMAL MONTAÑA DE GAIRÍA
CALDERA DE GAIRIA NARURAL MONUMENT

8 MONUMENTO NATURAL DE LOS CUCHILLOS DE VIGAN
NATURDENKMAL CUCHILLOS DE VIGÁN
CUCHILLOS DE VIGAN NATURAL MONUMENT

9 MONUMENTO NATURAL DE MONTAÑA CARDON
NATURDENKMAL MONTAÑA CARDON
MONTAÑA CARDON NATURAL MONUMENT

10 MONUMENTO NATURAL DE AJUI
NATURDENKMAL AJUI
AJUI NATURAL MONUMENT

11 SITIO DE INTERES CIENTIFICO DE LA PLAYA DEL MATORRAL
WISSENSCHAFTLICH INTERESSANTER ORT PLAYA DEL MATORRAL
PLAYA DEL MATORRAL PLACE OF SCIENTIFIC INTEREST

12 PAISAJE PROTEGIDO DEL MALPAIS GRANDE
LANDSCHAFTSSCHUTZGEBIET MALPAIS GRANDE
MALPAIS GRANDE PROTECTED LANDSCAPE

13 PAISAJE PROTEGIDO DE VALLEBRON
LANDSCHAFTSSCHUTZGEBIET VALLEBRON
VALLEBRON PROTECTED LANDSCAPE

FUERTEVENTURA
N
Carretera Principal
Main Road
Landstraße 1. Ordnung
Carretera Secundaria
Secondary Road
Landstraße 2. Ordnung
Camino de Tierra
Carriageway
Unbefestigte Piste
Puerto de Pesca
Fishing Port
Fischereihafen
Puerto de Pasajeros
Port
Passagierhafen
Palmeral
Palm tree field
Palmenhain
Windsurfing
Windsurfing
Windsurfen
Vista Panorámica
View Point
Aussichtspunkt
Aeropuerto
Airport
Flughafen
Puerto Deportivo
Yachting Harbour
Jachthafen
Castillo
Castle
Festung
Faro
Lighthouse
Leuchtturm
Molino
Mill
Windmühle
Las Salinas
Playa del Muellito
Puerto de la Torre
Playa de Leandro
Ensenada de Pozo Negro
Pozo Negro
Ensenada de Toneles
Punta Gorda
Ensenada de Jacomar
Punta de Gran Valle
Playa de los James
La Entallada
Las Playitas
Cala de Las Playas
Gran Tarajal
de Puerto del Rosario
Playa de Agando
a Morro Jable
Playa de Giniginámar
Giniginámar
Tarajalejo
Playa de Tarajalejo
La Lajita
Playa de la Lajita
Playa de la Jaqueta
Costa Calma
Playa de Sotavento de Jandia
Boca de Mal Nombre
Boca de Esquinzo
Playa de Butihondo
Morro Jable
Playa del Matorral
de Gran Tarajal y Puerto del Rosario
Playa de Juan Gómez
Punta Jandia
Puertito de la Cruz
Caleta de la Madera
Punta Pesebre
Roque del Moro
Cofete
Zarza
Playa de Cofete
El Islote
Playa de Barlovento de Jandia
Punta de Playa Larga
Agua Tres Piedras
Agua Liques
Istmo de la Pared
La Pared
Playa de la Pared
Playa del Viejo Rey
Puerto Nuevo
Playa de Ugán
Punta de las Goteras
Playa de Amanay
Las Salinas
Punta del Peñón Blanco
Playa de Garcey
Playa de la Solapa
Punta de la Nao
Ajuí
La Vega de Río Palmas
Mézquez
Agua de Bueyes
Casillas de Morales
Toto
Tiscamanita
PAJARA
TUINEJE
Las Casitas
Casas de La Florida
Catalina García
Juan Gopar
Tesejerague
Cardón
3
6
7
8
11
12

HISTORY AND CULTURE

Prehistory. Neither the archaeology nor anthropology of the Canary Islands is clear despite their having been extensively studied. Unexcavated archaeological sites, lost artefacts, buried settlements and the reuse of old materials in new building phases are some of the factors which prevent us from obtaining in-depth knowledge of our past.

In any case the basic questions - Who were they? Were did they come from? When did they arrive? have been asked for many years now, since the 16th and 17th centuries when the first General Histories were written by Alonso Espinosa, Leonardo Torriani and Abreu Galindo. In his 18th century "Historia Natural de Canarias", J. de Viera y Clavijo attempted to provide a portrait of the island's first inhabitants. In his view, they were the descendants of the inhabitants of Atlantis, a theory which aimed to ascribe lofty origins to the island's original inhabitants.

It would seem that the succeeding waves of new arrivals took place over a relatively long period of time. Certain authors, such as Schwidetzsky, basing their views on anthropological studies of European populations, hold that settlement of the Canaries took place in two main phases. The first corresponded to a Cro-Magnon type human - a broad, robust face and a more rounded body, and the second corresponding with a more Mediterranean-like human - taller, slimmer with a longer more delicate face. This second wave were more highly evolved in cultural terms.

Traces of human settlement in the lava fields.

Prehistoric house in Barranco de la Cueva.

Doorway of volcanic lava.

The Cro-Magnons from Mechta-el-Arbi (in Constantine) or Afalu-bu-Rhummel (in Beni-Segual, on the Kabilia coast near Algiers) both in Algeria, are the so-called "Mechta men" or "Mechta-Afalu" who arrived in the Canary Islands at the end of the Neolithic Age. Their primitive craft would have been subject to the ocean currents and trade winds, preventing them from returning to where they had come from or travelling to other islands. We do not know how many of these settlers arrived or whether they arrived in one or more phases. In any case, they must have come in sufficient numbers to allow later reproduction of the community.

Other groups of Africans followed these "Mechta men", many of whom would have been proto-Mediterraneans or their descendants, during the first Bronze Age. This leads to the conclusion that the newcomers originated in the Berber lands of the Sahara and the Maghreb. The remains of semi-excavated houses similar to those found in Morocco and certain areas of Tunisia have been found, in addition to idols bearing a marked similarity with zoomorphic Saharan sculptures. Pottery finds, both in shape and ornamentation, are also strikingly similar to similar items found in the northern part of Africa.

The gradual desertification of the Sahara, along with political problems in the north of Africa during the first millennium BC, were it would seem the most likely reasons for this migration. The lack of finds of human remains has prevented scientific analysis of the physical characteristics of these early inhabitants - the Majos. Nor do we have any written chronicle providing us with this information.

The language of the Majos was, according to Bontier and Le Verrier, "very similar to that of the Canaries". It would have been within the framework of the Lybico-Berber linguistic belt extending along the north coast of Africa. However, a number of alphabet-like inscriptions have also been found which add to the mystery. These show a Latin influence and are exclusive to Lanzarote and Fuerteventura. No equivalents have been found in other areas of the Archipelago.

Indeed, the term "guanches" generally applied to the entire aboriginal population of the Archipelago is nothing but a reflection of a mixture of peoples - Mediterranean, North-African and Negroid, originating from different immigratory waves and reaching the islands perhaps 1,000 years BC

According to most accounts, when the Europeans reached the Canaries, the island population was unable to sail and had no boats.

The natives, as a rule, were cave dwellers, living in existing caverns on the slopes of the mountains and ravines. When this was not possible they dug a hut into the ground and lined the walls with stones. On reaching ground level, they made a roof using flagstones that were then covered with doughy earth to prevent leakage. These dwellings have been given the name of "casas hondas" ("deep houses"). Another type of construction was the oval-floored house, with large walls of round stone; which were called "tagóror". They may have been centres of worship. Some had stone floors, with arched doorways made from natural stone like the walls. The inhabitants lived in settlements such as that of Barranco de la Cueva, which is the largest and best preserved on the island. In other settlements, such as Los Caserones in the north, the houses were more conical in shape.

The island was first known by the name of Erbania, in reference to the wall that divided the territory, separating two indigenous tribes. The wall was situated on Istmo de la Pared, with the followers of Ayoze living on the northern side, and those of Guize, to the south. The term Mahorata or Maxorata was originally used to refer to the caves in which the aborigines (the "majos" or "mahod") lived and was later applied by chroniclers to refer to the entire territory.

It was in 1339 that Angelino Dulcet produced a map in which the island's name was given as Forte Ventura.

We know even less about the lifestyle and customs of the Guanches, though there are many theories that are based on fantasy more than science.

We know that the settlements were small in size, that it was not a polyandrous society (women did not have more than one male partner in each family group), and that transhumant farming took place. Their diet was essentially one of meat and shellfish, mostly limpets and mussels. The remains of their shellfish meals gave rise to the shell beds on the north coast. There was also a dominant or privileged class, called "Altai" or "Altiai" - the courageous ones.

Descriptions provided by the island's first conquistadors confirm that the Guanches were of low stature and stocky body build, with fair and curly hair. It was a population in decline. Due to their Berber origins, they had some knowledge of metal working and other skills which the scant resources available in Fuerteventura prevented them from developing.

The Conquest. Ever since the days of the Ancient World and up until the conquest of the island, indeed, even since mythological times, people had known about the Canary Islands. Such authors as Plutarch, Pliny the Elder and Ptolemaeus had all written of them.

According to some historians, the Arabs came to the islands in 956 and 1013 BC In any case, it is certain that both Muslims and Christians knew of the islands, and may indeed have visited them more often than our evidence suggests.

Another certainty is that after the end of the 13th century, the islands began to be visited with relative frequency by European sailors. Technical advances such as the invention of the compass and compass cards, in addition to better

Desfiladero del Mal Paso, in the La Peña ravine. Béthencourt entered through here during conquest of the island.

Casa de los Coroneles.

and more accurate maps and charts, led to an extraordinary flourishing of sea travel.

The period prior to the definitive conquest of the island was marked by two distinct phases. The first began in at the start of the 14th century and lasted for its duration. In this phase the visits by Europeans were sporadic and they came in search of slaves. The second phase took place in the 15th century and was to lead to the conquest and occupation of the islands and their incorporation into the Crown of Castile.

If we limit ourselves strictly to Fuerteventura, then it was Majorcans, Catalonians, Basques, Andalusians, and Portuguese, together with Genoese, Florentines, Venetians and other merchants and navigators from the Italian peninsula, who carried out expeditions prior to the conquest. These trips originated from the Iberian peninsula. The Italian and Portuguese expeditions in 1341 and the Majorcan one in 1342 all made use of the charts of the islands drawn up by Dulcet in 1339.

The 1341 Portuguese expedition, funded by King Alonso IV of Portugal and led by Nicoloso da Recco and Angiolino de Teggia was to give rise to a dispute between the Portuguese monarch and King Alfonso XI of Castile, who laid claim to the Canaries on the basis that they had belonged to Mauritania Tingitana under the control of the Visigoth monarchy, which the Castilian kings saw themselves as heirs to.

The same argument was employed by the Castilian king in 1344, when Pope Clement VI, in his bull "Tua Devotionis Sinceritas", named Don Luis de la Cerda - the self-proclaimed Prince of Fortune - as sovereign of the islands.

In 1352, the missionary expedition of Arnau Roger took place under the auspices of the king of Aragon. The missionaries that came to the island were Catalans, as were the merchants, such as Bernat Merman and Pere d'Estrada. The Catalan presence did not last long, and had ended shortly after 1387, the year Peter the Ceremonious died.

According to Pedro de Ayala in his chronicle of King Don Enrique III, in 1393 another entirely Andalusian-Castilian expedition was organised, which captured 150 aborigines for the slave market. Seeing this, many greedy

knights urged the King of Castile to grant permission for a conquest. One of these greedy soldiers was Jean de Béthencourt who, with another Frenchman Gadifer de La Salle, was to proclaim himself as the island's first conqueror. From then on, our knowledge of events becomes more confusing and vague.

Between 1391 and 1393, relations between Castile and Genoa soured due to naval and commercial rivalry over North Africa and the Canary Islands, once their economic and strategic importance and become evident.

Béthencourt's announcement of his conquest of the island, and the recognition of his exclusive right to its ownership by the King of Castile on 10 January, 1403, ended all spontaneous travel to the islands.

Jean de Béthencourt arrived at the eastern islands with twenty-three Normans and forty Andalusians in 1402. He entered Fuerteventura through Puerto de la Peña (Ajuí). Not only had he come in search of slaves, but he also had the idea of establishing military domination, for which purpose he had contracted Gadifer de La Salle.

The first census (between 1440 and 1450) revealed a population of 1,200, living mostly in and around the capital of the island.

The first settlements were situated in the Valley of Vega de Río Palmas and Betancuria. At the beginning of the 16th century new settlements were to appear in the Valle de Santa Inés, Llanos de la Concepción, and later, Antigua.

The conquest of the island lasted three years. However, it was not a period of continuous war, but rather accommodation and coexistence between conquerors and conquered.

The Lord of the Islands of Fuerteventura and Lanzarote, a position established in 1404, was a subject of the king of Castile. He was also owner of all the islands conquered and responsible for their defence.

Between the 15th and 16th centuries much of the land was sold, and it was not until the beginning of the 17th century that the Dominion of Fuerteventura was consolidated with the House of Arias and Saavedra.

The Franciscan friary of San Buenaventura (1414)

Pirate attacks took place during the 16th and 17th centuries, mainly motivated by the search for water and meat. Only one, in 1593, led by Xabán de Arráez, could be considered as an invasion. The island was under Berber rule for six months, by the end of which, the capital and its archives had been destroyed. Against the threat of such attacks, defensive structures were built such as those in Barranco de la Torre, El Cotillo, and Caleta de Fustes.

By the mid 17th century, we have evidence that the island's occupation had spread as far as the El Cotillo area, which would seem to indicate that it was by then in use as a fishing shelter. There is also evidence that there was a chapel in Vallebrón.

Over this period, the island had a population of between 1,503 and 3,341.

Philip II forced the Lord to accept that defence should be in the hands of a professional soldier with military training. This was the origin of the position of the Sergeant Major, who assumed responsibility for defence in the absence of the Lord. Government of other matters was delegated to the Cabildo or Island Council.

From the Sergeant Major, Military government was gradually transferred to the Major Commandant, to the Lieutenant Colonel, and finally to the Colonel. The latter gradually began to take on such functions as Post and Tax Officer, and took advantage of his position to acquire almost the entire northern part of the island.

The permanent absence of the Lord benefited the Colonel, who went so far as to make his position hereditary, becoming for all practical purposes the new Lord of the Island, under a regime known as Coronelato (Colonelship).

The disappearance of the lordship and colonelship began with the Constitution of Las Cortes de Cádiz and finally ended with the establishment of eight town councils on the island.

Modern History. For various reasons, whether political, economic or natural, Fuerteventura passed through a decadent period during the 16th and 17th centuries.

The natural catastrophes occurring in the island's recent history have had a profound effect on the island and its people. Drought, and storms, in addition to other provoked catastrophes such as epidemics, wars and invasions, all gave rise to hunger and emigration.

After two centuries of futile attempts at to consolidate a grain farming economy, economic renewal began at the end of the 18th and beginning of the 19th centuries with production of barrilla, cochinilla, orchilla, cascarilla and jicanejo (aicán).

Salzkraut (Mesembryanthemum crystallinum)

Orchilla (Roccella s/p).

The barrilla, like the cosco, was collected and left to dry before being burned, and thereby converted into compact black lumps: the so-called "barilla stone". In this state it was exported to England and used as a source for caustic soda in the making of soap.

The cochinilla (Dactylocus cocus) popularly known as "grana", is a parasitic insect that lives on the leaves of the prickly pear-cactus. The parasite was placed in a little bag over the stalk until it laid its eggs. This occurred in early spring. They were then collected in summer or even autumn, using a special scoop, in a difficult and painstaking task. Once collected, they were left to dry before final sieving. From the cochinilla, a valuable carmine dye was extracted, which was used to make lipstick and dye.

It was introduced from Mexico in 1826 by Royal Order. An acclimatisation centre was also established and classes were organised teaching the inhabitants how to cultivate them. Grants were provided to encourage the industry and this led to increased production. The high point of the industry, in terms of grants, international success and prise rises, was in the 1851 - 1852 period. However, discovery in 1862 of a new artificial colouring, aniline, led to a slump in prices.

The orchilla is a lichen species of the Roccella family which grows on the northern side of volcanic rocks, sopping up the humidity carried by the ocean breezes. It is bush-like, rounded and elongated (5-10cm) in appearance, and grey-blue in colour. It was collected and exported for the manufacture of dyes.

The cascarilla or limpet shell was used in the manufacture of chicken feed. Many of the island's shellsbanks - archaelogical sites of great interest - ended their days in a chicken pen feeding trough.

Finally, aicán, which is a lichen that grows on volcanic rocks, was used as a dye and for tanning hides.

Grain farming has made its mark on the present landscape through the numerous mills (molinos and molinas) scattered throughout the island.

Trade in lime and salt was added to cereal farming in the 19th century. This led to a proliferation of lime-kilns and saltworks (which will be described in the section on architecture).

During the 18th and 19th centuries, there were also periods of famine caused by extreme droughts. This led to widespread emigration, first to Las Palmas and Tenerife, and later to South America -Montevideo, Buenos Aires, Mexico, Venezuela, and Cuba.

The island's inhabitants continued to plough the soil, piling up rows of stones to prevent erosion, and terracing the land for cultivation, until after the Spanish Civil War.

Molino de Tefía.

Promotion of commercial trade took place and Puerto Cabras became the island capital. The north of the island was developed and the first mercantile port was established. The port of Gran Tarajal was also opened to facilitate the production and export of tomatoes.

The advent of the modern tourist phenomenon has reversed demographic tendencies. However, this brings with it a risk that the island may lose aspects of its identity. Traditional forms of inland employment, for example, have been displaced as people move towards the coast.

On 16 August, 1982, a Statute of Autonomy was established for the Canary Islands.

Through this law, the Canary Islands Autonomous Government took on responsibility for the defence of their interests, balanced development of the islands and fostering of solidarity between all the inhabitants.

The Canary Islands are made up of the seven main islands of El Hierro, Fuerteventura, Gran Canaria, La Gomera, Lanzarote, La Palma, and Tenerife, and the smaller islands of Alegranza, La Graciosa, Montaña Clara and Roque del Este and Roque del Oeste, administered by Lanzarote, and Lobos, which is administered by Fuerteventura.

The role of capital is shared between Santa Cruz de Tenerife - seat of Parliament - and Las Palmas de Gran Canaria.

Each island is organised into municipalities. The islands are governed by the Island Councils or Cabildos and the municipalities by the Municipal Councils. The Canary Islands government is responsible for co-ordinating the activities of the Island Councils.

Fuerteventura is comprised of six municipalities: Puerto del Rosario, La Oliva, Betancuria, Antigua, Tuineje and Pájara. Each municipality is made up of a number of districts, often quite distant from the municipal capital and, ironically, larger. This is the case of Gran Tarajal in Tuineje, Morro Jable in Pájara, and Corralejo in La Oliva.

The Economy. Before the conquest of the island by the Europeans, Fuerteventura's economy was mainly based on agriculture and livestock raising, along with harvesting of wild vegetables, fishing and shellfish gathering. The cultivated crops included barley, wheat and broad beans. The main livestock were goats, sheep and pigs. Evidence of the fishing and shellfishing that took place can be found in the extensive shell banks found all around the island

After the conquest of the island, the best conditions for agriculture were found around the inland settlements. Agriculture has generally been of a subsistence type, with occasional periods of progress seeing an increase in exportation: first with cereals, later with the cochinea, and more recently, with tomatoes.

At present, one of the main items cultivated for export is Aloe Vera, which is used in the cosmetics and pharmaceutics industries.

The great number of windmills on the island were built to pump the irrigation water for agricultural uses, a function now performed by electric and diesel motors.

Traditional agriculture was marked by two fundamental features: "gavias", of African origin, were small earth dams in the beds of the ravines, built as a terraced series, such that when one part is filled with water, it overflows into the next lower part, and so on. The retained water is absorbed by the earth in a phase referred to as "la bebida" or "drinking". Then, the soil is planted. The other characteristic is the salinity of the wells, which limits irrigated crops to two species: tomatoes and alfalfa, the latter being most tolerant to the salinity of the water. Nowadays, fresh water is used for irrigation of tomato plantations.

Livestock raising played a secondary role due to the scarcity of pasture land, and although it is currently in decline, considerable numbers of livestock are still kept to produce the excellent "Majero" cheese. Now that an official "denominación de origen" (an official stamp of quality) has been created for the island's cheeses, promising export markets are appearing.

A curious point of interest is the classification of the island's goats into three distinct categories, depending on the degree of control exercised over them by the goatherds:

- Jairas: these are the domesticated goats that the goatherd keeps at home to provide milk for his family. These are chosen from among the goats that give the greatest quantity of milk.

- Cabras de ganado: these are the goats that are actually tended by the goatherd, usually

Windmill for pumping water.

A bardino.

Goats graze all over Fuerteventura.

between 40 and 100. Every day he takes them from his farm to the pasture and brings them back at nightfall. There he milks them, later to make cheese with the milk.

- Cabras de costa: are the goats allowed to roam through uncultivated fields. These goats live in a semi-wild state in the "costas" (uncultivated land), where they reproduce and raise their young. The kids are classified as "cabritos" - from the time they are born until they are weaned and "baifos" - from the time they are weaned until the appearance of their horns. The goats are properly marked.

These semi-wild goats are brought together from time to time in what is known as an "apañada"; the goatherds meet before dawn at the edge or "raya" of the uncultivated areas or "costas", with their bardinos, or herding dogs. They agree on a plan for the roundup and take the goats back to the corral or "gambuesa". There they are divided up and the ones that are unmarked are duly marked. Some of them are again set free until the next "apañada". Such days are usually holidays and festivals.

Fishing, like agriculture, has also been of minor importance.

In Fuerteventura, fishing is of the traditional as opposed to industrial variety. The fleet is composed of small boats of up to two tons, crewed by two men. Generally they do not fish for periods in excess of 24 hours, at the end of which, the boats are left on the beach or anchored, depending on the weather.

In recent years the sector has undergone changes and is increasingly oriented towards the tourist industry. There has been a proliferation of deep-sea game fishing craft and a major competition takes place in Gran Tarajal each year.

The different ports fulfil different functions and are as follows:

Commercial ports: Puerto del Rosario and Gran Tarajal.

Fishing/leisure/ferry ports: Corralejo and Morro Jable.

Leisure ports: Caleta de Fustes.

Anchorages and dry docks: El Cotillo, Puerto de la Peña, La Lajita, Tarajalejo, Ginijinámar.

Fishing ports: Puerto de la Cruz, Pozo Negro, El Matorral, Puerto Lajas, El Jablito, Majanicho and Los Molinos.

The fishermen are organised into three "cofradías" or local guilds:

Corralejo - from Puerto de la Peña to Pozo Negro.

Gran Tarajal - Pozo Negro to La Pared.

Morro Jable - from La Pared to Puerto de la Peña.

«Nasas» on the beach at Ajuí.

There are four areas whose physical and biological characteristics make them especially important for the reproduction of certain species of fish. These are:

Punto de Tostón - between Islote Rodrigo and Punta Gorda.

La Entallada - from Punta Gorda to El Jablito.

La Jaqueta - from Punta Gerepe to Punta Paloma.

Punta Jandía - from Punta del Tigre to Puertito de la Cruz.

These areas will serve as natural fish reserves for dispersion of the species into surrounding waters.

The following is a summary of some of the techniques and types of fishing, beginning with the simplest:

- Fija.- an iron or steel rod with one or several points, used for catching octopus in the low-tide pools.

- Gueldera or Tarralla.- a large sack of fabric or metallic mesh supported by a metal hoop. Hung from a thick flexible pole using four strands of fishing lines, it is used for catching the "carnada" (bait): bogues, chub mackerels, and blue jack mackerels. This technique is employed from a boat.

- Nasa.- a trap of iron framework and metallic fabric. Those used in the island are round and vary in size according to their use. They have two openings or "nasillos", and a door for removing the fish. It is the device most commonly employed in these waters, and in fact, the most deadly. Nasas are effective at any depth and on any type of sea bed.

A traditional «falúa» used for coastal fishing.

- Chinchorro. - a type of trawling net comprised of two bands and a pouch that is controlled from the shore or from a boat anchored close to the shore. It is used for catching small fish of different species: bogues, chubs, sardines, striped red mullet, axillary sea bream, and others.

- Albacora (Little Tunny) Fishing.- is done during the daytime, far from the coast, usually near the "veril" (verge of the coastal platform). The boats either anchor or drift, depending on the tides and the location of the fishes. Lines known as "liñas" are used. They are of varying thickness and the last few yards are of steel cable with straight, robust hooks. Usually two or three of these lines are used at a time. Live bait is preferred. When a fish bites. the other lines are pulled up so as to avoid entanglements. Depending on the weight of the fish, the struggle to land it may take longer or shorter. Fish of several hundred kilos could well take more than two hours. For skipsack tunas and little tunnys weighing less than 20 kg., a rod with thick casting-line is used.

- Cherne (Stone Bass) Fishing.- this is one of the most difficult techniques and requires great knowledge and skill. The stone bass are generally found in the deeper waters of the "veril", at depths of between 400 and 800m The wire line used is strong, resistant, and therefore instantly transmits the feel of a bite to the fisherman. At the end of the line there is a stretch (one and a half "brazas" long) of thick nylon line with several large hooks. At the extreme end of this, there is a heavy lead or iron which is known as a "chumbo". For the fishermen, a "braza" is the distance between the tips of the extended arms. Fresh bogues or chub mackerels are used for bait, three or four on each hook, strung through the eyes. The stone bass can weigh up to 90 kilos. Imperial blackfish, hake and several species of sharks are also caught.

- Rock Fishing. By "rock" here is meant a specific point of the sea bed where the concentration of fish is highest and where the fishing is good. The boat can either be anchored or adrift. The apparatus used is called a "cordel".

The advent of tourism has changed the island's economy. The beach at Morro Jable.

Networking!

It consists in a strong nylon line more or less thick and several hooks and a lead or steel weight at the end. With the boat anchored the usual catch consists of common sea bream, pink dentex, dusky groupers, comb groupers and other rock fish; at night - conger eels, rabbitfish and oilfish. The bait is either what is known as white bait (squid) or sardines, bogues, chub mackerels, and gueldes. On the Barlovento (windward) sea, octopus is also used as bait. With the boat drifting, the main catch is usually combers and, at night, axillary sea breams.

- Palangre (boulter).- Basically, it involves a line - the "mother" line - from which a number of hooks hang at even spaces. At the ends of the mother line, there are two floats or buoys and two "potalas" or "rociegas" (which are used to anchor the line and draw it in). The combined buoy-line-anchor is called a "cabecera". It is used for fishing common sea bream, pink dentex and fiddle fish. Squid is the preferred bait, and it is used at night-time when the large sea bream come onto the shelf in search of food. The palangres are normally left out for two to six hours. The hooks used are small and straight, so that they can be swallowed whole and become lodged deep in the fish's throat or stomach. It is usually pulled in at sunrise, and in addition to the fish mentioned above, large numbers of common stingray and skate are also caught. This technique is not normally employed at depths in excess of 100 m.

- Parrot fish fishing.- This is done from a boat, over a clear bottom, using a rod. The boats used are short and often motor-less. The rod tip incorporates a section made from male goatskin, which increases its sensitivity. The main bait used is "carnada de vieja" (Xantho porssa) - a small crab, or another, larger crab, known as a "juyón".

The present-day tourist boom has seen the introduction of numerous new activities into Fuerteventura's economic panorama, generating new patterns of work and the presence of emigrant labour from all over Europe.

A windfarm near Costa Calma.

Customs. Fuerteventura has retained may of its traditions and customs. While the younger generations may live a less traditional lifestyle, the trace of the past can still be found throughout the island.

One of the main traits of the Majorero is his hospitality. As Ramón F. Castañeyra observed long ago: "the Majorero is courteous, obsequious, and a friend to the stranger, to whom he will offer his best bed, and his finest victuals". He continues, "his character is not aggressive... he is reasonable and peaceful. Debaters and gifted in the defence in matters of rights and law."

Traditional weddings still take place and pregnant women drink lavender water if there is thought to be a risk of miscarriage. Between birth and christening, a candle is lit every night, and the old nursery rhymes and lullabies are still remembered.

The old superstitions about witches still exist and the saints are invoked against them, vows are made and amulets worn.

The superstitions include the following:

- "It is a sin to kill a wagtail - the "little birds of Santa María" - for when the Virgin fled with her Son to Egypt, they erased the trail left by their feet so that Romans wouldn't find them".

- "The devil appeared in the village of Betancuria; The townsfolk joined together and tied him up. They forced him to carry the stones that were used to build the church, then they let him go. The rope that was used to bind him can still be seen in the Cave of San Diego." The Cave of San Diego is in the chapel of San Diego, in front of the ruins of a Franciscan monastery. Miraculous powers are attributed to the floor of the cave.

Many of the superstitions involve death or disaster, although some others are of a more benign nature.

- Misfortune will befall a house with an "aromero" (acacia).

- The presence of quail in the countryside is a good omen.

- The presence of hoopoes in the fields is an omen for a good harvest and happiness.

- White spots on the fingernails are the sign of a liar.

- Warts are the sign of a person who spends the night counting stars.

Keeping things in order. Las Playitas.

Chatting away the time.

On the plain of Mafasca (Antigua) and the surrounding area there is a strange natural phenomenon known as the light of Mafasca which has given rise to a number of folk stories, in which the phenomenon is attributed to the suffering souls. Tales and legends of witches are also associated with the Tindaya mountain, which is shrouded in a veil of magic.

Vows made include offerings of candles and wax figures. In some chapels, such as Virgen de la Peña, wax hands, arms, heads and animals can still be seen. In general, it is women who make these offerings.

Vows and promises are made to San Roque to stave off contagious diseases; to San Isidro for a good harvest; to San Andrés for rain; to Santa Lucía to heal afflictions of the eyes; to the Virgen del Buen Viaje for those going to sea, and so on and so forth to a total of some 20 saints.

The most popular amulets are the bones of San Diego or pieces of his clothing, or remains of the apostles.

The folk culture of the islanders also includes a good number of games and sports, such as Canary Island wrestling, currently popular enough to fill stadiums and have a number of star competitors. There are ball games such as "pelota a mano", "juego de bote" or "pelotamano"; and "pina", a game of Arab origin. The card games include "seiseño", "tute", "politana", "primera", "burro", "brisca" and "truco".

These traditional folk pastimes have survived the passing of time through their innate attraction for the inhabitants and, in more recent times, through the work of a number of researchers.

A grounded ship was the centre of much conversation.

Many of these games were originally played among the island's aborigine population, and this explains their link with economic matters. In general, they resemble the games and pastimes found elsewhere. Others, were almost certainly introduced by the Europeans, such is the case of "pelotamano", which was in all likelihood, introduced to the island in the early 15th century by the conquistadors.

The origins of most of the table games, guessing and dice games, lie in magic rituals. They are games of chance, in which the outcome does not depend exclusively on the player's performance. Most of them are part of a ritual accounting for the presence of a distinguished personage, ceremonial celebrations, rules, and so on.

The traditional games and sports also reflect the values of a people. They change and evolve over time as do all aspects of culture; some however, remain largely unchanged. They also serve to show who is the best, and therefore serve as mechanisms for gaining prestige. They involve challenge and determination, and as a result some of them have even marked out an economic niche for themselves. The best example of this is Canary Wrestling, some of whose stars have attained considerable social status.

At present, after a long battle to survive, traditional games and pastimes seem to be in fashion again. Let us wish many years of life and health to the traditional games!

In the realm of purely festive diversions, there are numerous traditional dances and songs: the dance of "el perrito", the fandango - which is different from the Andalusian version; "aires de Lima" - folk-songs to be sung at any type of dance by two people alternately, usually a man and a woman; the "Juan Periñal" dance; the "folías", a popular song and dance throughout the islands, and "seguidillas", "isas", "malagueñas" and the "polka majorera" complete with burlesque and often spicy personal references.

The folk medicine of Fuerteventura provides a great number of remedies for common ailments. Bleeding of the arms, herbal teas, massages with "mejunjes" (medicinal mixtures), herbal extracts, enemas and purgatives, poultices and other concoctions are employed to cure pneumonia, muscular weakness, nasal congestion, diarrhoea, eye problems, headache, burns, and many other ailments. Special mention is deserved by the use of "tripa de cucas" (cockroach guts) for the relief of toothache. According to tradition, if the pain persists it is because the patient did not wish to take the cockroach guts that were prescribed.

Prophecies or predictions on the basis of such signs as rain, the sunset, goat's excrement or the female camel's hair still draw the attention of the older inhabitants in Fuerteventura. The following are samples:

If the skies are clear at sunrise on 21 September, San Mateo's Day, it will turn out to be a good year; if it is misty or foggy, a bad one. If in autumn, goat excrement breaks up easily, it will be a good year; if hard, a bad one. If on 24 June, San Juan's Day, sunrise coincides with high tide, a good omen; with low tide, a very bad one.

To conclude two Majorero proverbs: "In September, let he who has bread start sowing", and "a camel never sees his own hump". However, perhaps it is best to leave the traditional riddle that follows unanswered:

In a valley not altogether flat
There are two crystalline springs;
The gardener is not too pleased
When the flow dries up.
What are they?

The boats decked out for the festival of Carmen. Morro Jable.

Fiestas. The present-day festivals are, to a greater or lesser extent, derived from the island's ancient practices and culture. Every town has its annual festivities in honour of the patron saint in whose name the local church was built.

These festivities are events in which public participation, spontaneity and generosity, not to mention simplicity, are centre-stage. People buy new clothes, house fronts and doors are given a coat of paint, the plazas are decorated with flowers and flags, oxen are slaughtered, "verbenas" or street parties are organised, sweet dishes are prepared, along with community meals; there are bonfires and the people throng down to the beach. These are just the occasion on which to hear the "parrandas" singing the traditional songs, "isas", "folías", "seguidillas" and "malagueñas".

Most fiestas involve the celebration of a sung mass with a sermon and procession. The church is adorned with flowers and banners. The fiestas generally last for three days. The most important local ones are mentioned in the chapter on the different municipalities. Here we describe the fiestas that are common to all the island.

The Carnival is undoubtedly one of the most popular of the islands' festivals. It takes place during the month of February in all the towns of the island, the main celebrations being those of Puerto del Rosario and Gran Tarajal. Carnival is the celebration of the transvestites, the liberation of desire, a return to childhood, a journey into old age and a lampooning of public figures. People are grouped into "murgas" and satirical songs are sung by all.

The "romería" (pilgrimage) is another type of fiesta, more religious in nature. There are two especially popular romerías: the "Virgen de la Peña" through Mal Paso - Vega de Río Palmas, which takes place every third Saturday of September, and brings together pilgrims from all parts of the island to offer their vows to the Virgen de la Peña, patron saint of Fuerteventura;

The second great romería is the El Tanquito romería, in El Cardón, which is truly unique. The townspeople carry a figure of the Virgin on their shoulders to the top of the mountain. The route is a long one and there are frequent stops to regain strength and energy,

Then they sail along the coast, with the statue of the Virgin.

accompanied with music and dance. At the summit, the Virgin is placed in a grotto and the fiesta continues with a traditional barbecue.

Another main fiesta, in the maritime tradition, is the celebration of the Virgen del Carmen which takes place in Corralejo and Morro Jable. Every 16 of July, the townspeople parade the figure of their Virgin along the shore.

The fiesta Jurada de San Miguel, on 13 October of each year, is in fulfilment of a vow made to the saint following a victory over the British troops who invaded Fuerteventura in 1740. The festival began in 1946 on the initiative of the parish priest of Tuineje. After a lull, it was resumed in 1974 and has been going strong since. From 8 to 13 October a series of performances take place in Tarajalejo, Tiscamanita, Gran Tarajal, and mainly, in Tuineje, with staged recreations of the historical events on which the festival is based.

Like all features of folk culture in today's world, the traditional fiestas are coming under the effects of the cultural standardisation that is sweeping the modern world. Instead of the intimate celebrations of a local community, today's festivals see the influx of huge numbers of tourists seeking entertainment and leisure based around the traditional festivities.

As mentioned earlier, these festivities are all closely related to traditional music and dance. Today, folklore experts search for, study and adapt old songs from before and after the time of the island's conquest.

We know very little of the music of the aborigines. Their instruments were rudimentary: primitive percussion instruments for the hands and feet, necklaces made of sea snail and limpet shells, the latter being known as "cuchars de lapa". With conquest of the island by the Europeans, new instruments were introduced - tambourines, triangles, castanets, scrapers, and so on. At a later stage, these were added to by guitars, mandolins, lutes, and "timples". The latter is the Canary folk instrument par excellence. Also known as the "camellito", it is an adaptation of the Iberian guitar with a long narrow body which is convex on the under side. The number of strings and the exact tuning may vary from island to island. Another characteristic instrument is the accordion or "forito".

With the advent of the new instruments, new songs and dances appeared:

- "Isas" - these are relatively recent songs (1940s-1950s) derived from earlier forms. In the dance, the men carry a stick from which hangs a traditional woollen bag.

- "Malagueñas" - sad songs usually on a theme of love for one's motherland or laments for a dead mother. It is derived from the Andalusian fandango and is unusual in that there is a chorus verse after each solo verse. The dance is basically for soloists. The man leads the woman, or women through its choreography.

- "Folías" - one of the best-loved. It derived from a baroque dance which became popular in the 17th and 18th centuries. It is comprised of two sections of eight bars each, which are repeated with variations. The dance is for a single dancer with castanets.

- "Seguidillas" - as a literary stanza, the form has existed since the 15th century, and in musical form since the epoch of the Catholic Monarchs. They are sung by a number of soloists - as many as required - and there are no instrumental interludes.

- "Sorondongos" - their origins date back to a children's game in the 16th century. The dance comprises circles of men and women and individuals who take turns and dance simple steps in the centre.

- "Sarandas" - a relatively recent creation (1970s) which has consolidated its position in Majorero tradition. The dance is around a "saranda" (winnow) held aloft by a group of men

The festivals bring an opportunity to sing the traditional songs.

and women in order and imitating the movements of workers in the fields.

Apart from the festivities of patron saints, the other setting for expressions of folk culture were the "bailes de candil" - literally "oil-lamp dances". These were lively dances held in rooms lit up by an oil lamp, with a small musical group supplying the music. When the room was particularly small, the men had to pay a fee called a "taifa" to dance, and the dances were then known as "bailes de taifa". Sometimes, they were known as "bailes de gobernadores" because a person known as the "governor", armed with a stick, was entrusted with the task of ensuring that no fights broke out among the men. At a given moment of the dance, the music would stop and "teases" or "challenges" would be sung between men and women, or on occasion, only among the men.

These traditional dances began to fade out in the 1940s.

The folk dances that have survived is thanks to the groups that have made an effort to preserve Fuerteventura's folk culture and traditions. Without them, it would all be just another part of history.

Arts and Crafts. Crafts production in Fuerteventura is somewhat less than prolific, but one can still see craftsmen working in traditional forms that have been handed down from generation to generation. These are an important part of the island's cultural heritage and as such, should be cherished.

At present, there is a handcrafts centre open to the public in La Oliva. Every year in May there is a handcraft fair in Antigua which draws craftspeople from all over the Canary Islands.

The embroidery of the island is renowned; a variety of styles and techniques are practised in different regions. The finished product can be bought in Lajares, Tuineje, Tindaya, Tefía and Puerto del Rosario, and in many cases visitors can see the craft workers at their craft

The island's fretwork and lace are also widely admired. The fretwork from Antigua, Puerto del Rosario and Tindaya is especially famous.

In the past, handmade pottery was a feature of daily life on the islands. Ceramic items were used for cooking, storage and transport of food as described by Abreu Galindo "...they ate from sun-baked earthenware vessels, like large pots..."

As in the other islands, before invention of the potter's wheel, the pottery in Fuerteventura was handmade and of low quality. It was generally made from a clay base, to which they sometimes added other materials, small stones and shells, and on which they placed cylindrical layers until reaching the desired height. The product was then smoothed down with a stone to achieve a shine and consistency before final baking. Any decoration employed was by cutting and tended to be simple, consisting of parallel, horizontal, vertical, triangular and broken lines.

There are still a number of traditional pottery workshops scattered throughout the island, such as that of our friend Silverio in Rosa del Taro, between Triquivijate and Puerto del Rosario, which keep the Fuerteventura pottery tradition alive.

Tindaya and Llanos de la Concepción are also known for high-quality handcraft products made from palm leaves.

In Tuineje there has been a revival of interest in textiles woven on traditional looms. Virgin wool, yarn, and camel hair are used.

Finally, Betancuria and Vega de Río Palmas are known for basket-weaving and wickerwork.

Jandía lighthouse.

Architecture. The inhabitants of Fuerteventura over the centuries have left their mark on the island's landscape, constituting a unique patrimony which, in addition to major monuments, also includes numerous spaces which provide an insight into life as lived on the island in the past.

The historical buildings can be classified into four groups:

- Religious buildings
- Unique buildings
- Industrial buildings
- Residential buildings

Religious buildings. These include 12 churches and some 20 chapels scattered throughout the island.

The churches were built as the various urban centres developed. They reflect the styles in vogue at the time of their building and also the high points of the economic history of the area in question. Many of them were built over long periods of time, and most have required rebuilding at some stage whether due to fires, pirate attacks or simple renovation and extension. All of them feature unique characteristics, such as the classic façade of the church of Vega de Río Palmas, the majestic façade work of the triple-nave Church of La Candelaria in La Oliva, the markedly Aztec-like doorway of the church in Pájara, the stonework doorway of Casillas del Angel, and the bell-towers of Antigua, Tetir and San Miguel de Tuineje.

Tetir cemetery.

A chapel bell-tower.

Most of the chapels were built during the 18th century. Erected by the villagers themselves, the construction and material used are accordingly simple.

They usually comprise a rectangular nave and a sacristy attached to one of the sides. Some of the corners include stonework of volcanic rock. The V-shaped roofs are usually of wood covered with adobe and, in some cases, festooned with a row of tiles. The austere interior usually includes a small pulpit, a baptismal font, an altarpiece and some sacred paintings.

The Franciscan friary of San Buenaventura.

Unique buildings. Buildings which, for their location, use or origin are considered unique: cemeteries, forts, lighthouses and warehouses ("cillas" and "taros").

Until the beginning of the 19th century, the deceased were buried beneath the church at Betancuria. Later, municipal cemeteries were built, today totalling twelve. Except for those along neo-classical lines (as in Pájara or Puerto del Rosario) and the more recently built, the rest are traditional in style: whitewashed stone walls with small pyramidal ornamental flourishes, rectangular bases and the chapel behind.

Forts were built for the defence of the ports. By the mid 17th century, Caleta de Fustes, El Tostón and Tarajalejo were indispensable for transport and trade.

In 1741, the forts of El Toston and Caleta de Fustes were built as strategic points. These two-storied structures have a circular base and are shaped like a truncated cone. The upper floor, with its gentle dome roof, supported the aljibe or water cistern. The lower

El Tostón fort. El Cotillo.

La Entallada lighthouse.

El Cotillo lighthouse.

level contained the living quarters and service rooms. Both floors were insulated by air chambers. These forts have been restored and are in good condition.

The island's four lighthouses date from the beginning of the 20th century. They generally have a square base, central patio and a lateral tower (Lobos, El Tostón and Punta de Jandía); however, the lighthouse of La Entallada in Tuineje is unique in all the islands, both for its structure and the detail of its finish. There is a new lighthouse at Morro Jable and another in Puerto del Rosario.

Two types of warehouses exist:

The "cillas" are warehouses that were built by the church to store the tithes and its own harvests. Various structures of this type existed in Tetir, Tiscamanita and La Oliva, the latter being the only one in good condition.

The "taros" are rural structures in the tradition of the local agrarian economy.

They were made of stone, mud and lime, and served for the storage of fodder, cheese and farm tools. They usually have a circular base, three to five metres in diameter, and three to six metres high, depending on the number of floors. At present there are six such taros, in Lajares, La Matilla, Pájara, and Ampuyenta.

Industrial buildings. Buildings linked to the island's economic activity: "tenerías" (tanneries), "tahonas" (bakeries), "molinos" (mills), "hornos de cal" (lime kilns) and "salinas" (saltworks).

The tanneries included all the spaces and tanks required for the tanning of hides. Only the Mézquez tannery remains.

Given the island's grain and livestock history, the machinery for grinding the grain was highly important.

The earliest implement was the hand-mill, of Roman origin, simple to use and construct. It comprises two circular equal parts, one on top of the other, known as "muelas".

Lime kiln. Barranco de la Torre.

The Tiscamanita Windmill.

El Carmen saltfields.

There was also a long shaft ending in a point which could be inserted in a number of openings on the upper millstone and which served to provide support to the movement. The millstone is of basalt. This mill was used by the family to grind the "gofío" eaten on a daily basis.

These hand-mills led to the tahona, which was used by a larger community. The tahona was part of the country house and is an important feature in the island's agricultural tradition. A forerunner to the mill, the first tahonas were built at the beginning of the 18th

Crystallised salt.

century. They were worked by a group of villagers ("moliendo a bote") or, more usually, by a camel harnessed to the "almijarra", the piece which drove the complicated mechanism of millstones, shaft bearings, wheel-spindles and so on. The grain entered through a hopper. The structure has a square base, walls made of stone, mud, and lime, a one or two-sheet roof covered with adobe. A great number of tahonas remain.

Mills began to appear on the island in the 17th and 18th centuries. Replacing the tahonas, they are unmistakable features of the landscape. There are two types of mills: the "molino macho" (literally, male mill) and the "molina" (literally, female mill).

Because of their great historic importance, the chapter on "Touring Routes" includes a "mills tour". This trip will, it is hoped, provide visitors with an insight into an important part of the island's cultural heritage and traditions. Descriptions of the different mills are given along with details of their locations.

Lime kilns were first built on the island in the 17th century. The lime was used to purify rain water, to make whitewash and for building. Distribution of the kilns was determined by the location of the "calicheras" (caliche beds) and trading patterns. Most of them have a circular base, some are rectangular, others a mixture of both. The construction varied depending on whether the kiln burned coal or wood. The smaller ones burned wood and were circular, between three and six metres in diameter, with only one opening. Some were semi-underground, or located in the ravines to facilitate introduction of the raw material and extraction of the finished product. The coal-burning kilns generally produced lime for export, were rectangular-shaped, from four to sixteen metres in length and more than ten metres high, with as many as four openings of different sizes. There are 149 lime kilns on the island, that at Rosa de la Arena in Puerto del Rosario being especially noteworthy.

The lime production technique basically remained unaltered from the time of the Romans until the advent of cement in the 1950s. Caliche,

Salt

the raw material, came mainly from beds formed by fossil beaches, from which it was extracted in the form of chips. The chips were then further broken down, cleansed of impurities and placed in the kiln. In the kiln they settled over an iron grille, under which a fire was lit using dry furze. Layers of wood or charcoal were then intermingled with the caliche, and the overall cone was covered over and topped off with a large stone. It took between three and four days for the entire column to burn. A change in the colour of the smoke announced completion of the process. The quicklime obtained was then quenched with salt water and finally sieved to remove "palomillas" or impurities

Last but not least, we come to the ingenious saltworks. They were designed to obtain salt through the evaporation of seawater by the sun.

The story of salt is the story of the great civilisations. The earliest reference to it dates from ancient China (2205-2197 BC). Tax on salt was, until relatively recent times, a basic source of income for the state. And despite all the advances made in chemistry and refrigeration, salt is still a universal product with an almost magical power of survival.

We have very little written testimony of salt production among the island's early inhabitants. We know that they obtained it

<table>
<tr><th colspan="4">Technology of the salt</th></tr>
<tr><td>Harvesting</td><td colspan="3">Harvest the salt deposited in a natural way: in shore-cliffs, small lagoons, coastal or big continental deposits.</td></tr>
<tr><td>Mining Industry</td><td colspan="3">Obtaining of salt by means of mining uses of salt gem</td></tr>
<tr><td>Graduation and boiling</td><td colspan="3">Obtaining of salt from heating of the sea water or the brine in recipients of different materials using some class of fuel until producing the evaporation of the water and the precipitation of the salt.</td></tr>
<tr><td rowspan="6">Concentration and evaporation
(it only uses solar energy)</td><td>Extensive Saline</td><td colspan="2">The crystallization doesn't appear divided. They are usually interior, of spring.</td></tr>
<tr><td rowspan="5">Intensive Saline
(all the Canary saline)</td><td>Naturals</td><td>High tide puddles on muddy or rocky shore-cliff lands. They only possess the still.</td></tr>
<tr><td rowspan="4">Artificials
(they consist of: reception and impulsion, still, watering systems and crystallization)</td><td>Primitive on rock. They have two recipients: the still and the crystallization vessel or surge. The reception system is by means of an overflow connected to the still. They are circular and closed with stone walls and mud or mortar of lime.</td></tr>
<tr><td>Ancient of mud. Generally settled on accumulations of alluvial character. The walls of the still and pipes are made of stones caught with mortar of lime and waterproofed with mud or lime. The bottoms are of rammed mud. Saline of El Río (Lanzarote).</td></tr>
<tr><td>Ancient of mortar of lime. Masonry factory of basaltic stones caught with mortar of lime, as much the walls as the bottoms.</td></tr>
<tr><td>New of mud with stone linings. Typical of Lanzarote and Furteventura. Except for those from El Río. All the other ones are of this type. The position of the still is invested. Now they are in the interior and the water moves toward the cuts closer to the coast. The cuts are composed: divided by interior "pareditas" (stop). They are built with stone and mud, being only used the lime for the impermeabilization inside the pipes. The bottoms are of rammed mud.</td></tr>
</table>

directly from the pools left by the high tide and used it to salt their fish. In some locations, these pools were modified so as to obtain a higher yield. It is quite possible then, that before conquest of the island by the Europeans, saltworks of some sort were exploited by the indigenous population.

After the conquest, the natural salt pools passed into the ownership of the lords, although the locals had the right to make use of them. In 1525, the Crown granted this right of use to private owners, and in 1605, the existing saltworks were included among the monopolies of the Royal Treasury. It was after this period that the first real development of the island's salt resources began to take place. The next great upsurge in the salt industry took place between 1910 and 1930, especially in Lanzarote, thanks to the salted food industry.

The lifting of salt taxes at the beginning of the 20th century and the application of new

Doorway in Betancuria.

A balcony on a «señorial» building. Betancuria.

technology in the design of the saltworks - with linings of stone and lime - greatly aided this upsurge.

The crisis in the salted food industry and the invention of refrigeration techniques heralded the end of the salt boom, the only remaining vestige of which today are the now restored Carmen saltfields, which provide a unique insight into the island's historical and cultural heritage.

The seawater (4% salt) is captured by "tomaderos" or intake pools. This water is then channelled to the "cocederos" or salt pools, where it remains for a period of between 15 and 20 days, during which it is transferred from one recipient to another. This process is known as "curtido" and the aim is to achieve a salt concentration of up to 20%. It is then transferred through pipes to what are known as "tajos" or crystallisers. The saltworks are protected from the wind by walls made of volcanic rock. The salt remains in the tajos for some 20 days, until total precipitation takes place. Between 12 and 14 salt "harvests" take place in a year, between the months of March and October. The salt obtained is accumulated on the edge of the pools, where it is left to drain until the next harvest takes place.

The saltworks tend to be located in low-lying, flat areas bathed by high tides and open to sunlight. These natural conditions are aided by a layout which ensures that gravity plays a role in channelling the flow of the salt water, or by other mechanisms such as windmills and, sluices.

Apart from the natural saltworks, such as those at Aguas Verdes, most are artificial: El Matorral in Morro Jable, Isla de Lobos, Puerto Cabras in Puerto del Rosario (no longer in existence) and those of El Carmen in Antigua, the only ones to have been fully restored and worth a visit.

Residential buildings. There are two general categories: the traditional houses reflecting folk architecture of the island and the "señorial" houses - the homes of the nobles.

Rural house with volcanic rock lintel.

Traditional houses. La Alcogida. Tefia.

The great virtue of the traditional house is that it blends into the landscape and is adapted to the prevailing geographical and climatic conditions: constant wind and scarcity of water. They were located according to the proximity of the cistern or well, and sheltered from the wind. The final finish of the stone walls developed over time, first with the appearance of white borders around the doors and windows, and later with the introduction of high quality whitewash.

The best examples of folk architecture are found in the rural areas, in many of which the houses have barely changed over the years. The floor-plan is basically a four-sided one, within which another central square forms a patio. The various rooms, the number of which varies, open onto the patio.

The building materials used are those closest to hand and needing least processing. At first, houses were built in dry stone, then with stones and mud, later using with lime mortar, and finally using blocks of volcanic sand and cement. The roofs were usually V-shaped and made of "pírganos" (the central section of the palm leaf), hawthorn or "bobos" (Nicotiana glauca), and lats and panels of tea tree. The floors are "torta" - a mixture of "tegue" (clay) and "granzón" (straw). Externally the walls are given an "a cuchara" treatment (covered without smoothing off) which leaves a rugged surface to which the successive layers of whitewash are applied.

Lintel made of compressed volcanic ash.

The buildings were generally located close to whatever productive activities took

Traditional house next to Casa de los Coroneles, La Oliva. In the background: Montaña del Frontón.

Window of the same house.

place and are usually facing south-east. The most important area of the house was the kitchen, and the main facility the "aljibe" or cistern. Other items such as presses, cellars, the threshing area, animal pens, the drying area, plus the type of wood used and the number of windows gave each house its own characteristic air. The kitchen also functioned as a sleeping area and living room. For cooking, a "tenique" (hot stone) was first used, followed by the "reverbero" (a small alcohol stove) and, later, simple burners and oil-cookers were introduced.

The kitchens also house the ovens, which are basically cubic or cylindrical in shape and the top part being finished off with a spherical cap. Inside this cap there is a hollow space and it is covered with a special light volcanic stone, which maintains the heat.

The high point of the kitchens are the fireplaces, with their markedly Moorish influence.

The cisterns were built in low areas or hollows. The water first passed through a

Casa Isaítas. A «señorial» house restored for use as a rural hotel and an excellent restaurant. Pájara.

sieve made of furze which helped to remove sediments. When the cistern ran dry, they made use of the well.

The noble's home, clearly colonial in influence (Portuguese, Andalusian, or Castilian), is associated with ecclesiastic and administrative power (Betancuria, Antigua, Pájara), military (La Oliva), and later, wealthy farmers (Tuineje, Antigua, Tetir, Casillas).

Food. For reasons of climate and custom, Fuerteventura's cuisine tends to be quite plain. The main ingredients used are fish and island produce flavoured with special sauces and seasoning.

Most dishes are served with a centuries old island-accompaniment known as gofio, made from toasted grain flour. "Gofio amasado" is made from a doughy mixture of gofio, water, milk, broth, potatoes, honey or wine, served in a "zurrón" (leather or fabric bag) or casserole dish. Most gofio is made with a mixture of toasted wheat and barley flour. Wheat and barley were once the island's most important crops. There is also "gofio de millo", made from a toasted course cornflour.

Other dishes, served as appetisers, include:

"Papas arrugadas", small unpeeled potatoes steam cooked with plenty of salt and served with "mojo picón" hot sauce.

Majorero goat's cheese, perhaps the island's product best known for its quality.

Limpets served in «mojo verde»

Salemas in preparation.

Pejines, tiny fish of the sardine and anchovy family, dried in the sun, then baked, grilled, or cooked in an alcohol flame.

Shellfish; mussels, limpets and "burgaos" (Gibbula candei) steamed or pickled.

Crustaceans; spider-crab, crabs, shrimps, lobster and goose barnacles.

Octopus and cuttlefish, roasted.

"Viejas jareadas a la brasa", fish cleaned, filleted, and dried for 24 hours, then grilled over hot coals.

Soups, including those made from fish, potato with scrambled egg, and corn. There are also bread, courgette, and pea soups.

The tastiest dishes include: Canary Island stew, with garbanzo and kidney beans, and limpet, rabbit, hen, tongue, roast meat, cod, and fried fish, chicken, raw grouper, octopus, pigeon and kid dishes.

Kid, for example, is cooked as follows: the kid is cut up, dipped in brine and then fried. Meanwhile, a number of slices of bread are also fried along with a few smashed cloves of garlic. Water is added, a little vinegar, ground black peppers and chopped parsley. The mixture is brought to the boil and the fried kid is then added. Simply delicious!

Other traditional and delicious dishes include: baifo meat dipped in "mojo picón" (hot sauce) and fried; and rabbit in "salmorejo" sauce.

Fish abound in the island's cuisine. Some examples:

"Sancocho": sea bass salted and poached with potatoes and sweet potatoes, served with hot sauce.

Windsurfing at Jandía.

"Cazón con mojo": dogfish, a tender and succulent species, eaten fresh or dried with sauce.

"Viejas" (parrot-fish), "salmonetes" (striped red mullet), "cabrillas" (combers), "sargos" (white sea breams), "morenas" (fangtooth morays), etc.: fried or boiled and marinated with oil and vinegar, are some of the island's delicious fish dishes.

There can be no better way to round off a good meal than with a good dessert. The choices include ground rice puddings, milk desserts, sweet pineapple, almonds, "suspiros", and cheese pudding - made as follows: to a pound of goat's cheese you add a pound of butter, a pound of sugar, twelve yolks and six whites of eggs. Beat the mixture well, place in a cake tin, and into the oven.

Nothing remains but to order a good wine and enjoy your meal!

La Barca beach, Jandía. Great windsurfing both for beginners and experts.

SPORTS

All of the town councils have a sports centre, however modest, permitting a wide range of leisure activities. Many hotels, and apartment complexes also have sports facilities.

Open air sports include hunting, fishing, diving and windsurfing. We will provide all the necessary information for each of these.

Fishing. Given that intensive professional fishing already takes place, sports fishing must take every precaution not to overfish or damage marine resources. For this reason, there are special rules which must be observed.

Canary Island mussel fishing: the minimum size permitted is 7 cm long. The mussel may not be collected between 1 April and 30 June, and 1 September and 30 November.

Coastal fishing: no form of trawling or net fishing, especially trammel nets, is allowed.

Recreational coastal fishing:

Underwater fishing. no night fishing; the use of canisters, means of propulsion or electric, electronic or explosive mechanisms is prohibited. Fishing rigs or traps may not be used. Maximum permitted catch per person per day is 5 kg; groups of more than five people are limited to a maximum catch of 25 kg. Underwater fishing may not take place within 250 metres of other people, including those fishing on the surface.

Underwater fishing is allowed in the following areas:

- El Jablito, from Puntilla to Cabo del Agua.
- La Entallada, from Punta Toneles to Gran Tarajal.
- Cofete, from Roque del Moro to Punta de Amanay.
- Los Molinos, from Punta Catetones Mansos to Playa de Tebeto

Recreational surface fishing. a maximum of three hooks may be used at one time, and only two rods may be used per person. Maximum catch per person per day may not exceed 4 kg, or only one fish of greater weight. For groups of more than four. The maximum permitted catch is 16 kg.

Recreational deep-sea fishing. may not be carried out within half a mile of the professional fishing craft. A maximum of three fish may be caught.

In all the three cases mentioned above, the catch must be for one's own use and may not be sold. More than 10 kg of fish may not be transported from one island to another per person.

Remember that for recreational fishing, a special licence is required from the Dirección General de Pesca.

There are tables with details of minimum sizes of fish permitted for each species. They may be found at the headquarters of the fishermen's associations in each port.

Hunting. The Regional Hunting Council has also established regulations for hunting. These include the following:

Hunting seasons.

Hunting of rabbit with dogs and ferrets may take place from 2 to 16 August.

Hunting of all permitted species with dogs, ferrets and rifles may take place from 23 August to 29 November.

Hunting may take place on Sundays, national holidays, and the day of the Virgen de la Peña. The following species of bird may be hunted: common partridge, Barbary partridge, quail, turtledove, rock dove, rabbit and Barbary squirrel.

Birds of the fringillidae family may not be hunted (goldfinch, siskin, greenfinch, linnet, serin, wild canary, gold-breasted waxbill) anywhere on the island; however, ferrets may be used for rabbit hunting.

Hunting with compressed air guns, calibre 22 rifles or firearms with more than two cartridges is prohibited. Possession and use of cartridges of more than 2.5 grams weight is also prohibited.

Possession and use of lures or decoys making use of recorded sounds, live decoys

A boat which ran aground in the sands of Jandía.

which have been blinded or mutilated, and use of partridge decoys is prohibited.

Hunting is not allowed in wetland areas (ponds, reservoirs, saltmarshes and saltfields, etc.) or within 50 metres of such areas.

Hunting is not allowed using motorised vehicles or boats.

Use of birds of prey for hunting is not allowed.

With the aim of protecting certain animal species or for reasons of safety, hunting of all species of birds is prohibited in the following areas: Madre del Agua refuge (Betancuria), Reservas Biológicas de Lajales (La Oliva), Laguna de Tesjuate (Puerto del Rosario) and the Jandía isthmus. Rabbit hunting with dogs is permitted, subject to authorisation. Hunting is prohibited in the Castillo de Lara reserve (Betancuria) and in the Parque Natural de las Dunas de Corralejo and Isla de Lobos (La Oliva).

Hunting out of season is considered as a serious offence.

Groups are limited to a maximum of five hunters. One hunter may have four dogs and groups may have a maximum of ten dogs. Maximum quarry is four rabbits and four partridges per hunter per day and 12 pieces per group. For fowl, a maximum of five turtle doves and 15 pigeons per hunter per day, is permitted. Eggs and young animals count equally for the calculation of maximum permitted quarry. Possession or sale of protected species is prohibited.

Surf at Caleta del Hierro. Majanicho. La Oliva.

Diving. For those wishing to discover the secrets and mysteries of the marine depths, below there are some guidelines concerning the sea and the facilities available to divers.

The main factors underlying the marine conditions are the Canary current and the deep, nutrient-rich waters off the coasts of Africa. These determine both the temperature and salinity of the waters, which in turn, determine its characteristic flora and fauna.

The tides are oceanic and there are two high tides and two low tides per lunar day. The greatest tidal amplitude is 2.7 m, which occurs with the spring tides in February and September.

The conditions are perfect for diving, with temperatures between 17 and 23° C, acceptable levels of visibility and transparency and a total lack of dangerous species.

In Fuerteventura, as in all the Canary Islands, there is a clear distinction between two marine areas: windward (Barlovento) and leeward (Sotavento).

The windward zones, exposed to the prevailing trade winds for the greater part of the year, generally present unfavourable diving conditions. However, they tend to be richer in marine species, not so intensively fished, and more interesting for the recreational diver with an abundance of rocks, caves and ledges.

The windward zone in Fuerteventura extends from Corralejo to Punta Pesebre at the extreme end of the Jandía peninsula.

Any windward diving excursion should depart from the fishing village of Ajuí, in the coastal municipality of Pájara. Just off the beach of Ajuí, without any need for support boats, there is an area of great interest to divers with a rocky bed, immense caves, steep walls and passageways overflowing with large fish species.

By renting a fishing boat, you can access two of the most interesting diving spots in all of the islands: La Cresta and La Herradura. Only highly experienced divers should attempt these sites.

Other interesting dives are possible on the same coastline, a few hundred metres to the north, off the beach of El Jurao.

On the leeward, south-facing coast, the sea is relatively calm throughout the year. The best diving facilities are located here in the major tourist centres - Corralejo, Caleta de Fustes, Costa Calma and Morro Jable.

In Corralejo, across from the island of Lobos, there is an extensive rocky area, the site of a future marine park, with many interesting features. In the central coast area, there is good diving at Las Salinas, especially at Jacomar, where the nearby volcano spilled its lava into the sea.

Another excellent internationally known diving area is at Morro Jable. It consists of a wall of sandstone full of crevices and caves, with the largest concentration of large fish of the entire island. Large groupers, yellow mouth barracudas, guelly jacks, fang tooth morays, and many other species accustomed to divers, guarantee an unforgettable underwater experience.

There is a decompression chamber in Jandía, should divers encounter any decompression problems.

Windsurfing. Given the sea and wind conditions, the coast of Fuerteventura is ideal for this sport. The best areas are off the shores of Jandía and Corralejo, which also boast the best facilities and the majority of windsurfing and sailing centres. Caleta de Fuste is another popular spot. For the more experienced, there are the windward shores, with high winds and seas, but no back-up facilities. You need your own equipment. The speed trials for the World Windsurfing Championships are held each year off the beaches of Jandía, during July and August.

PUERTO DEL ROSARIO

Up to the end of the 18th century, Puerto Cabras was a place where the goat farmers from Tetir and surrounding areas took their herds to graze and to drink from the many fresh water springs.

The history of Puerto Cabras begins with the building of sailor's huts alongside the goat pens. Around 1805, María Estrada, a native of Tetir, opened the first tavern in a stone house. From then on, the town continued to grow, helped by the barilla trade, which had reached major proportions by 1810.

In 1806, the town separated from the parish of Tetir and built an altar dedicated to the Virgen del Rosario.

The first part of the town was built in 1808, and this was followed by a period of hardship on the island. The town continued to grow and in 1835 it was officially recognised as a separate town to Tetir. Later, in 1860 it was made the official capital of the island because of its port, geographic location, and commercial importance.

The only products exported in those days were barilla, orchilla, cascarilla, jicanejo and *musgo* (moss). Between 1925 and 1926, Puerto Cabras absorbed the municipalities of Tetir and Casillas del Angel. It was given its present name of Puerto del Rosario in 1956.

Nowadays, in addition to the districts mentioned above, Puerto del Rosario, also includes El Matorral, El Time, Garcia Escámez, Guisguey, La Ampuyenta, La Asomada, La Matilla, Llanos de la Concepción, Puerto Lajas, Tefía and Tesjuante.

Puerto del Rosario has a total surface area of 290 km2, stretching from coast to coast and the town itself stands at an altitude of 16 m above sea level. The area's economy is based on services: transport, communications and trade; followed by construction and manufacturing. The importance of agriculture and fishing has now fallen both here and in the rest of the island.

The local fiestas are Fiesta de la Virgen del Rosario (1-7 October); and Ash Wednesday.

Of special interest are the following:

The village of La Alcogida, in Tefía, a representative sample of traditional island architecture: stone walls, footpaths, use of water resources in the traditional "gavia".

The parish church of Casillas del Ángel, dedicated to Santa Ana, with its original neo-colonial black stone tower; The parish church of Tetir, dedicated to Santo Domingo, rebuilt in the 18th and 19th centuries, and featuring a beautiful altarpiece and some fine paintings.

Also of interest are the chapels of La Matilla, La Asomada, El Time, Tefía, and San Pedro de Alcántara in Ampuyenta.

Tetir.

LA OLIVA

The archaeological remains found in this townland attest to the existence of a significant aboriginal population, and it may have been the seat of the ancient pre-Hispanic kingdom of Maxorata. The first western settlements date to the beginning of the 14th century.

As the population grew, a chapel was built and a friar was brought from the monastery of Betancuria to oversee it. In 1708 the parish church was built and in 1711, La Oliva was officially recognised as a separate town from Betancuria.

In 1708, the Militia Regiment and the post of Colonel - Governor at Arms - were created, and La Oliva was chosen as the permanent residence for the Colonels, thus making it, in effect, the military capital of Fuerteventura. The Colonel also had civil powers; powers that were taken away in 1859 when the Regiment was dissolved by the Ministry of the Army. From then on, the seat of military command was to be Puerto del Rosario.

La Oliva became an independent municipality in 1812.

Today it includes Corralejo, El Cotillo, La Caldereta, Lajares, El Roque, Tindaya, Vallebrón, and Villaverde.

The municipality has a total surface area of 356 km2, and the capital stands 219 m above sea level.

The economy is based primarily on tourism, the main tourist centre being Corralejo. Tourism is followed in importance by fishing and livestock raising.

The local holidays are Virgen de la Candelaria on 2 February, San José on 19 March and Virgen del Carmen on 16 July.

Among the places of interest are:

La Casa de los Coroneles (the Colonel's House) dating from the 18th century. It is a sober two-story building topped with three towers, and built on a square base around a central patio with pillars of tea tree wood. There are 365 windows, one for each day of the year. The facade has eight wooden baroque style balconies and oven the main entrance we find the coat of arms of the Cabrera family, the fifth Colonel, who had the building built.

The parish church of La Oliva, dedicated to the Virgen de la Candelaria, features three naves and a facade with bell-tower of black stone. Inside are a number of interesting paintings.

The archaeological remains of Tindaya, an aboriginal place of worship.

The Castle of El Tostón in El Cotillo, built between 1741 and 1743, is the best example of island defence from that epoch.

And, finally, on Montaña Quemada we find a homage to the hospitality of the island's people, who understood the Spanish thinker Miguel de Unamuno

Parish church of La Oliva.

Houses on the lava field. Lajares.

who proclaimed himself defender of this "Tierra sedienta" or thirsty land.

The area's geography is highly varied and representative of all the island's geological phases. We find material from the basal complex, wide plains, volcanic remnants, the volcanic spike of Tindaya, recent "malpaises", great U-shaped valleys and extensive dune fields forming a spectacular landscape topped off by marvellous beaches of white sand.

ANTIGUA

Antigua's existence dates from prior to 1485. At first, it was a settlement of Andalusian and Norman families who lived from farming and livestock raising. Although the chapel dedicated to the Virgen de Antigua was one of the first buildings on the island, the parish was part of Betancuria until 1784.

In 1812, it became a separate municipality, and was the island's capital for a short period. For geographic and economic reasons, it became the seat of the Subordinate Governing Council, a decision which caused public unrest requiring the intervention of the militia to restore order.

Nowadays Antigua includes: Agua de Bueyes, Caleta de Fuste, Las Pocetas, Las Salinas, Pozo Negro, Valles de Ortega, Triquivijate, and Casillas de Morales.

The municipality's total surface area is 250.5 km2 and the town itself stands at an elevation of 254 m above sea level.

The economy is based on agriculture, a mixture of dry and irrigated farming (greenhouses) and livestock raising, which is common throughout the entire area. Fishing is confined to the port of Pozo Negro. The most important tourist centre is Caleta de Fuste, which provides an ever increasing number of jobs and is becoming one of the island's most popular destinations.

The local holidays are Nuestra Señora de Antigua on 8 September, and Shrove Tuesday.

Among the points of interest are:

The church of Antigua (18th century), with its clear Franciscan influence;

The archaeological sites of Atalayita and Junquillo;

The chapel of Agua de Bueyes, dedicated to Nuestra Señora de Guadalupe, and that of Casillas de Morales; the castle of Caleta de Fuste, built in the 18th century; and the many characteristic mills that are such an integral part of the plain.

The highest geographical point is Morro Janana at 674 m, on the Betancuria massif, which descends to an inland plain which covers most of the territory. To the east, there are lines of U-shaped valleys which served to channel the lava from the Gairía and Los Toneles volcanoes. These lava flows formed the Malpaís Chico, and the Malpaís Grande, which stretches along the Pozo Negro canyon.

View of Antigua from Morro Velosa.

TUINEJE

There is a lack of reliable data on the history of this municipality, though human remains in Tuineje and Tarajal de Sancho suggest that the region was settled prehistorically. From after the conquest up to 1590, it was a district of Betancuria, with a church dedicated to San Miguel. In 1790, it separated from Betancuria and became a parish in its own right.

Between the 17th and 18th centuries, it saw more turbulent times with continuous attacks by pirates and privateers who entered by the inlet of Gran Tarajal. One such invasion by the English took place in 1740, and was repulsed by the islanders at the battles of Llanos de la Florida and Tamasite. Though greatly outnumbered and poorly armed, the islanders, through cleverness and the use of dromedaries, managed to defeat the invaders.

In 1872, Tuineje became a municipality. At present it includes the districts of Gran Tarajal, Tarajalejo, Giniginámar, Las Playitas, Tiscamanita, Tesejerague, La Florida, La Fuentita, Catalina García, Las Casitas and El Charco.

Gran Tarajal, uninhabited until the beginning of this century, is today the second largest town on the island, and surpasses Tuineje in importance, although administratively it is secondary to it.

Total surface area is 276 km2, and the capital stands at an elevation of 205 m above sea level.

The most outstanding feature of the area around Tuineje are its great plains furrowed by numerous gullies. The coast is cliff except at the mouths of the large valleys: Tarajalejo, Giniginámar, Gran Tarajal, Las Playitas and Gran Valle. Tuineje shares the Nature Park of Pozo Negro, with its volcanic peak of Caldera de los Arrabales, with Antigua. The Vigán massif, surrounded by the Gran Valle and Del Roque ravines, is another point of ecological interest and provides an attractive walk along the coast.

In addition to the general flora found throughout the island, there are abundant palm and tamarisk groves in these valleys.

The local economy is based on agriculture, livestock, fishing, and tourism. In agriculture the tomato is the mainstay, with greenhouses located in all the valleys. Livestock raising is important throughout the region and supports a lively cheese industry. Gran Tarajal, Las Playitas, Giniginámar and Tarajalejo are the main fishing and tourism centres.

The local architecture is also of interest with its characteristic lack of the colour white, serving to

Las Playitas.

camouflage the houses against outside attacks. Also of interest among the prehistoric ruins is the "tagoror", of the Tarajal de Sancho aboriginal settlement. Religious architecture is well represented by the church of San Marcos, in Tiscamanita, which has Franciscan influences, and the parish church of San Miguel, whose altarpiece depicts the battle of Tamasite.

Local holidays are: Nuestra Señora de la Candelaria on 2 February and the feast-day of San Miguel Arcángel on 13 October.

BETANCURIA

Betancuria was founded by and named after the island's conqueror in 1404. It was at one point the second largest city of the Canary Islands. The banner of the conquest is still preserved in the cathedral museum and flies every year to mark the anniversary of Fuerteventura's incorporation into Castile, which occurred on 14 June 1405, during the reign of Enrique III.

The church built to celebrate the conquest was recognised as a cathedral in 1424 by Pope Martin IV. An archbishop was appointed, but never came to live on the island.

Betancuria was to be attacked on many occasions by pirates, the majority Berbers, and the cruellest of all being Xabán de Arráez who destroyed the capital and its archives in 1593, leaving with six hundred Christian prisoners. In the same year work began on the rebuilding of the town and the cathedral, which was not completed until 1691.

In 1667 it became the seat of the territorial Lords, remaining so until 1836. Until 1837 it was inhabited by Franciscan monks who were to leave an important legacy.

Later it became the seat of the Island Council, which together with the Lords, governed the island until 1820.

In 1812, the Cortes de Cádiz established it as a municipality, and in 1834, it ceased to be the island capital.

Betancuria includes both Valle de Santa Inés and Vega de Río Palmas.

The municipality occupies a surface area of 104 km2, and the town itself stands 395 m above sea level.

This area is the most dramatic in terms of landscape. The highest elevation is the peak of Betancuria at 724 m, from which the massif obtains its name and which is mainly made up of material from the basal complex. Practically the entire municipality is part of the Betancuria Rural Park, offering beautiful landscapes and excellent views from Morro Velosa on the road to Antigua.

The economy is based on agriculture and livestock, concentrated around the main ravines of Vega de Río Palmas and Betancuria. Because the coast is not suitable for settlement there is no fishing activity and it is tourism, centred on the historic town

Betancuria cathedral.

centre, that has had most recent impact in economic terms.

Betancuria is Fuerteventura's richest town in terms of cultural heritage, being especially noted for the old historic town centre, which has been declared an official Historic and Artistic Centre. This includes: the Cathedral, on which a 100-year long period of rebuilding and renovation has left the stamp of several styles: Gothic, Baroque (Hispano-Moslem), Renaissance and *Mudéjar*. The interior includes neo-classical side altarpieces, and the main altar is gilded and painted in the Baroque style. There is a fine collection of paintings and figurines and a late 16th century silver monstrance, thought to be the oldest in the Canary Islands.

The ruins of the Franciscan friary of San Buenaventura, authorised by Pope Benedict XIII in 1414, was the first in the islands. It was renovated and extended in 1460. In the 19th century, the ceiling collapsed, and has since remained as seen today. The church was built alongside the chapel of San Diego de Alcalá, who was buried there.

The Sacred Art museum in the old Parochial House was restored in 1983 and features objects from all the churches of the island.

Finally, the Betancuria Museum houses items of ethnological interest.

Other points of interest outside the municipal capital are the chapels of Valle de Santa Inés and Virgen de la Peña in Vega de Río Palmas. The annual festival of the latter sees participation by the entire island, with a procession of religious floats and festive stops along the way. The Las Peñitas reservoir draws its water from this ravine, and overall, the area is one of the most beautiful landscapes of Fuerteventura.

The local fiestas are: San Buenaventura - patron saint of the island - on 14 July. Fiesta de la Virgen de la Peña - patronness of the island - on the third Saturday of September, and San Bartolomé's Day on 24 August.

PAJARA

This municipality, one of the island's oldest, was founded by goatherds and fishermen who settled during the 16th century. In 1708, a chapel was built to Nuestra Señora de la Regla, and was under the administrative jurisdiction of Betancuria. In 1711 Pájara became a parish in its own right. In 1729, the militia was established and sent in by the *Coronel* at La Oliva in protection against possible attacks.

The church in Pájara.

Pájara is renowned for its long golden sandy beaches. Sotavento de Jandía.

Pájara was officially recognised as a town in its own right in 1812 and nowadays includes El Cardón, Chilegua, La Lajita, Matas Blancas, El Puertito, Puerto de la Peña (Ajuí), Toto and Morro Jable. Well worth visiting is the parish church of Pájara, with its two naves and a noteworthy façade on the main one. It is classical in style: a triangular pediment, with rosewindow included, supported by pilasters. Of particular interest are the Aztec inspired decorative motifs; suns, snakes, lions, doves, etc..., the origin of which is not known.

The municipality of Pájara has a total surface area of 383.6 km2, and it is the municipal area with the longest shoreline in Spain (150 km, 70 of which are beach). The city stands at an elevation of 196 m. The majority of the inhabitants live along the Jandía coastline. The most outstanding geographical feature of the area are its beaches, on both the leeward and windward coasts. The highest elevation is Pico de la Zarza (807 m), for further information see "Touring Routes". The Istmo de la Pared divides the territory into two distinct areas: it is a 4 km-long sand strip stretching from coast to coast and covered in dunes.

Tourism is the mainstay of the local economy, Jandía being the island's main tourist centre. Agriculture is based on tomatoes and potatoes grown in greenhouses in the valleys and irrigated areas. Livestock raising is important, favoured by the large areas of unpopulated land, goats being the preferred stock. Finally, the fishing ports of Ajuí, La Lajita and Morro Jable also make an economic contribution, though not a major one.

The local holidays are: Nuestra Señora de la Regla on 2 June, and Nuestra Señora del Carmen on 16 July.

Barlovento de Jandía. Cofete.

BIBLIOGRAPHY

In this guide, we have attempted to cover the greatest possible number of topics of interest. However, due to limitations of space, it is not possible to cover all in great detail. For readers wishing to read more, we provide a bibliography below, classified by content.

We would like to thank the authors. Without them, this guide would not have been possible, since most of the titles cited were used as sources in writing it.

General:

- J. M. HERNANDEZ-RUBIO CISNEROS:

"Fuerteventura en la Naturaleza y la Historia de Canarias".

- J. M. HERNANDEZ-RUBIO CISNEROS: "Fuerteventura hasta la abolición de los Señoríos, 1477-1837".

- CABILDO INSULAR: "Plan Insular de Ordenación de Fuerteventura". 1992

Geology:

- CONSTANTINO CRIADO HERNANDEZ: "Evolución del relieve de Fuerteventura".

- ARANA, V. & CARRACEDO, J. C.: "Los Volcanes de las Islas Canarias II. Lanzarote y Fuerteventura". 1979. Editorial Rueda.

Ecology:

- Sets of illustrations on Fuerteventura:

a) Los Fósiles de Fuerteventura en su Historia Geológica.

b) La Fauna Jandiense.

c) Los Depósitos del Inicio del Pleioceno y sus Fósiles.

d) El Jandiense: Estratigrafía en Las Playitas.

e) El Erbaniense y su Fauna

- BRAMWELL, D & BRAMWELL, Z.: "Flores silvestres de las Islas Canarias". 1990. Editorial Rueda.

- KUNKEL, G.: "Flora de la Isla de Lobos". Cabildo Insular de Gran Canaria. 1970.

- MORENO, J. M.: "Guía de las aves de las Islas Canarias". 1988. Editorial Interinsular Canaria.

- MARTIN, P & CARDONA, A.: "Avifauna Canaria II. Aves de zonas bajas". Cabildo Insular de Fuerteventura.

- MIGUEL PIZARRO: "Peces de Fuerteventura". Consejería de Agricultura y Pesca.

- ALEJANDRO GONZALEZ MORALES: "Estructuras agrarias recientes de Fuerteventura".

History, Art, Archaeology, Language:

- Volumes I, II, III and IV: "Jornadas de estudio sobre Fuerteventura y Lanzarote".

- ANTONIO DE BETHENCOURT & AURINA RODRIGUEZ: "Ataques ingleses contra Fuerteventura. 1740".

Customs:

- RAMON F. CASTAÑEYRA: "Memoria sobre las costumbres de Fuerteventura". Prologue, introduction and notes by FRANCISCO NAVARRO ARTILES.

- PEDRO CARREÑO FUENTES. "Recetas culinarias".

- El Zurrón magazine

Economics, Politics:

- CARMELO DOMINGUEZ HORMIGA: "El Sector Primario en Fuerteventura. Canales de Comercialización".

- CARMELO DOMINGUEZ HORMIGA: "Política turística en Fuerteventura".

Poems:

- DOMINGO VELAZQUEZ: "Los caminos".

- MIGUEL DE UNAMUNO: "De Fuerteventura a París".

- DOMINGO VELAZQUEZ: "Poemas del sueño errante".

-M. HORMIGA: "Poemas de Pe a Paz".

- MAXIMILANO TRAPERO: "Romancero de Fuerteventura".

Humour:

- "La Batalla de Tamasite". El Cuchillete.

Maps.

- Mapa Geológico de España (1:50.000). Instituto Geológico y Minero de España. C.S.I.C.

- Cartografía Militar de España (Serie L, 1:50,000 Serie C, 1:100,000). Servicio Geográfico del Ejército

- Map RAI EDICIONES, 1:100,000

- Aerial photography (1 page): 1:100,000. RAI EDICIONES

TOURING ROUTES

The routes have been drawn up with the following considerations in mind: we have tried to put together a selection that includes all the diverse landscapes of the island, while at the same time, aiming to reduce the negative consequences of a lot of present-day tourism. The roads are not suitable for low-speed touring with frequent stops, and there are few properly sign-posted routes. Therefore, the bulk of present touring is along a limited number of tracks. One of the objectives of this section then, is to try to channel the demand more rationally, thus avoiding the environmental degradation arising from proliferation of innumerable paths.

From among the most travelled routes, we have chosen the ones that have the most to offer. When taken together, they offer a varied range of environments. Although road conditions permit a choice of different types of vehicles, we recommend four wheel drive, which is absolutely necessary in some cases. Bicycles can be used on most of the tracks and are appropriate for the distances involved, though certain routes should perhaps be divided into stages. Two of the routes (Lobos Island and Pico de la Zarza) can only be done on foot and the others are of different lengths and range which encourage the traveller to venture into hidden beauty spots.

Each route description is accompanied by a map and photos. Each of the texts is divided into a block of general information and a description. The suggested time for each route is based on a generous estimate, allowing time for additional walks, swimming, rest and refreshment along the way. The time needed and the estimated level of difficulty (low, medium, high) assume that a four-wheel drive vehicle will be used.

The descriptions include general information about the surrounding environment for each route, including landscape, geomorphology, flora and fauna. As far as possible, the use of complex scientific terminology has been avoided. However, frequent use is made of the local names of species, geological and topographic formations to give the reader a sense of the area's cultural wealth.

The sketches of the routes are based on aerial photography (1:25,000) and names and other data have been added. The scale of each map varies according to the space available, but is always as close as possible to 1 cm = 250 m.

From the outset of preparation of this guide, one of our primary concerns was preservation of the environment. The selection and planning of routes has been entirely subject to this criterion. In addition, we believe that the better something is known, the better it will be protected, and in the case of Fuerteventura, a great deal needs to be learned if we are to appreciate many of its landscapes.

The Canary Islands have environmental legislation. It is perhaps adequate but insufficiently implemented and developed. Some of this legislation is mentioned in other sections of this guide. One example is the law of Protected Natural Spaces, which lacks regulations for actual use of these areas. This lack of criteria led to problems when we were drawing up the routes. Apart from the laws and penalties that have been established, the best and perhaps the only guarantee of environmental protection is personal commitment. The precautions to be taken are minimal, in most cases obvious, and always based on common sense and civic responsibility. Going off the marked trails, altering the vegetation, disturbing the fauna, taking specimens, lighting fires and camping, hunting and fishing without the required permission and information (seasons, locations and species permitted) are all activities that common sense alone should discourage, without the need for laws and penalties.

In spite of the limited space available, we believe that this guide will prove useful, not only to our visitors from abroad, but also for residents who wish to deepen their understanding of their own island. We also hope it will serve as educational material in schools and will help to motivate and guide the reader in the adventure of discovering Fuerteventura. We recommend routes that will bring you into contact with the sweeping landscapes of the plains, the mountain massifs, the malpaíses, the ravines, and of course, the magnificent craggy coasts and golden sandy beaches.

Sun protection for the head and skin, adequate footwear, water and some provisions are all that is required to enjoy the tranquillity and leisure of each route. Our hope is that the readers will enjoy the following itineraries at least as much as we did in compiling them, and also that they will serve to heighten appreciation of the unique and serene beauty of the landscapes of Fuerteventura.

CORRALEJO - EL COTILLO

DISTANCE	DURATION	DIFFICULTY
26 km	3-4 hours	Low

Access: Track along the coastline, leaving from the extreme north of the village of Corralejo (Morro de La Vieja).

Description: This route runs along the north coast of Fuerteventura between Corralejo and El Cotillo. The landscape comprises chiefly "malpaís" (neither rock nor earth; lava covered in lichens), "jable" (dune fields) and sea are the main elements of the landscape.

The beginning of the route offers good views of the inland peaks of Bayuyo and Las Calderas, part of the volcanic group of Calderón Rondo, which in turn is part of a chain that ends with the island of Lobos. The track leads to the edge of the Caldera Encantada malpaís, a stretch of recent lava flows about 100 km2 in area, originating from the above mentioned volcanic group. The eruptions which formed the malpaís were "tranquil" ones taking place over an extended period of time. The eruptions from the horseshoe-shaped craters (and some few with perfectly rounded craters - Calderón Hondo, La Mancha, Lomo Blanco) gave rise to some black volcanic structures ("picones") and extensive quantities of lava which flowed from fissures at the bases of the cones. This volcanic period resulted in the extension of the island to the north, burying an ancient dune field which, like the present one, marked the former coastline. The fossilised interior of this former coastline juts out in places and is visible at the surface.

After passing a small stretch of sand on Morro de La Vieja, we come to a fringe of malpaís area devoid of rocks and walls. The track goes through an area of basaltic slag which shapes the characteristically irregular surface. Thorny shrubs, "tabaiba", furze, "matomoro" and thyme are some of the species that have managed to take root on the lava, but lichens are by far the most numerous type of malpaís vegetation. Depressions with clayey surface areas built up by runoff water support a thicker scrub dominated by "matomoros" and "rama". In the lowest areas, occasionally washed by the tides, the vegetation is richer, resembling that of a salt marsh.

Some five km from the beginning of the route, we reach the Caleta del Mejillón, an islet of organic sand on the coast, with "salados" growing up to a metre and a half high. Inland, the peaks of La Mancha, La Raya and Lomo Blanco form another volcanic cluster running northeast to southwest. Around these volcanoes, there are numerous "hornitos", structures formed by the expansion of gas as it bubbled through the lava.

Three km further on we come to the Caleta del Barco, an inlet with a large flooded area frequented by mud-dwelling birds (shanks, plover, sandpipers). This area, known as El Jablito, stretches over the malpaís to the foot of Lomo Blanco. To the extreme west of the sands lies Majanicho, a wide shallow bay of largely similar characteristics to those described above. Here, the track branches off, leaving Majanicho and crossing the malpaís in the direction of Lajares. However, our route follows the coast, crossing a muddy depression covered with "matomoro", "salado", and "siemprevivas".

In Caleta del Hierro, there are clusters of "seba", a marine plant that grows in underwater meadows, creating an environment favourable to the reproduction of fishes. The tides have a "pruning" effect and deposit the cuttings on the shore.

This particular species lends its name to the next bay, Caleta la Seba. The colour, texture and composition of the sand is different, a prelude to the landscape and environs of El Cotillo. The contrast between white sand and black lava combines with the graduated blue and green hues of the water to produce a startling effect. The plants growing on the sand, with "tabaibilla" and "salado" crowning the dunes, are another outstanding feature of this uniquely beautiful location.

Two km from Caleta la Seba, we arrive at Caleta las Salinas, and 1.5 km further on, Caleta de Beatriz. Inland, bordering on the sandy areas, there is "tabaiba dulce" ("sweet tabaiba") growing on the malpaís, which will be seen again in the vicinity of El Charcón.

The track leaves the malpaís at Punta Aguda and continues over sand. "Tabaibilla", "balancón", "uva de mar", "corazoncillos" and "salado" are some of the species typically found on this sandy terrain. Outstanding amongst these is "ononis", a small bush

of the pulse family with tiny dark green leaves and yellow flowers, greatly appreciated by the local livestock.

Around the lighthouse of El Tostón, there are several coves (Marrajo, Aduana, El Río, etc.) of great beauty which are ideal for a pleasant stroll. On the shore, among the "sebas" thrown up by the sea, one can frequently see limicolous birds (mud dwellers) feeding and resting. From the lighthouse, we take a paved road for about four km, which brings us through malpaís, sandy stretches and urban settlements on the outskirts of El Cotillo. Our route ends at the Marfolín beach. At one to two metres above sea level, you can easily observe a fossil beach 6,000 to 8,000 years old, partly destroyed and fragmented, which runs the length of the coast from Corralejo.

El Jablito.

LAJARES - CORRALEJO
(Malpaís de la Caldera Encantada)

DISTANCE	DURATION	DIFFICULTY
15 km	3-4 hours	Low

Access: Turnoff from the road that connects Lajares with the road running from La Oliva to Corralejo.

Description: The Caldera Encantada malpaís, or Bayuyo, (Recent Volcanism, series IV) was formed more than 8,000 years ago from a great number of eruption points, which extended along Fuerteventura's north coast. The recommended route covers part of the old "camino real" or king's highway, which connected the interior of the island with the port of Corralejo, running around the northeast - southwest volcanic chain that ends at Lobos Island.

We leave Lajares by the road that connects the town to the road running from La Oliva to Corralejo (km 32). Alongside the road, the fossilised sands indicate the position of the ancient coastline, with clay-filled depressions produced by incursions of the malpaís into the local drainage pattern.

Two km from Lajares, we take a turnoff to the left which leads into the malpaís and towards the volcano chain. We travel over some rugged basaltic outcrops extending over an area that was once pounded by the sea.

The track takes us to the base of Caldera Rebanada (253 m), a volcano in the middle of the chain, formed by the build-up of slack known as "picón". It is crescent shaped and open to the northwest. After passing around the southern side of the volcano, at a point where the track veers towards the north, we come to a track made by the Cabildo Insular (Island Council), the entrance to which is marked by large rocks.

This track provides a pleasant and leisurely trip of about one hour's duration through malpaís, and provides access to the peaks of Calderón Hondo and Montaña Colorada. A branch of this path leads to the summit of Calderón Honda (223 m), the only perfect crater of this group, formed by layers of solidified slag and ash. Another branch of the path connects with a trail leading to Montaña Colorada, a volcano similar to Caldera Rebanada in the extreme southwest of the chain.

Along the trail, one can observe the various plant species typical of the malpaís. Between lava flows there is open scrub, made up of "rama", thorny bushes, "verode" and a few "mimos" and prickly pear-cactuses. In places there are patches of "tabaiba dulce" and rocks covered by varied lichen flora, among which the green shades of the "aicán" predominate.

Lizards and birds are the most frequently seen fauna on the malpaís. Among the birds, the most common are the "zarzalero", "pájaro moro", pipit, and the kestrel - a bird of prey. In springtime, it is easy to spot the nests of Barbary partridges and their numerous chicks.

From Caldera Rebanada we follow the track to Corralejo, and come immediately to the volcanoes of Caldera Encantada and Las Calderas. Both have the characteristic horseshoe shape, open to the east. A little further on, the Bayuyo and San Rafael volcanoes mark the end of this volcanic chain. They are arranged like a pair of horseshoes opening

Malpaís. General view from Calderón Hondo volcano.

to the northeast and northwest respectively. To the northwest, from the extrusions of Las Calderas and Bayuyo, rise the so-called "hornitos" (ovens), protuberances formed by accumulation of gas bubbling below the lava.

From Bayuyo (269 m), the highest elevation of this group, there is a fine view of the other volcanoes whose lava formed the malpaís. About two km to the northwest, beyond the "hornitos" made by this lava, we can see another group of three volcanoes running parallel to the previous group. These are Lomo Blanco and La Mancha (two perfect cones), and between them the small La Raya. To the east of Bayuyo lies the isolated cone of El Cuervo, whose lava flows extend to the coast and are now covered by the Corralejo jable (dunes). In line with Bayuyo, La Caldera (110 m) is the volcano that marks the northeastern extreme of the chain which, beyond the shores of Fuerteventura, gave rise to the island of Lobos. Beyond the malpaís and jable, on the other side of the Bocaina Strait, we see the coast of Lanzarote, with the tourist resort of Playa Blanca, the plains of El Rubicón, and the Los Ajaches Massif.

The trail continues on to Hoya del Caballo, over an area where the malpaís progressively gives way to jable which finally covers the lava in the vicinity of Corralejo. The landscape has changed dramatically. Contrasting with the stillness of the lava, is the mobility of the dunes, among which thrive distinct species of plants adapted to the peculiarities of life in the shifting sands. Towards the coast, the horizon is dramatically shaped by the outline of large "balancones" anchored on dunes behind the shore.

3·GC·2·842

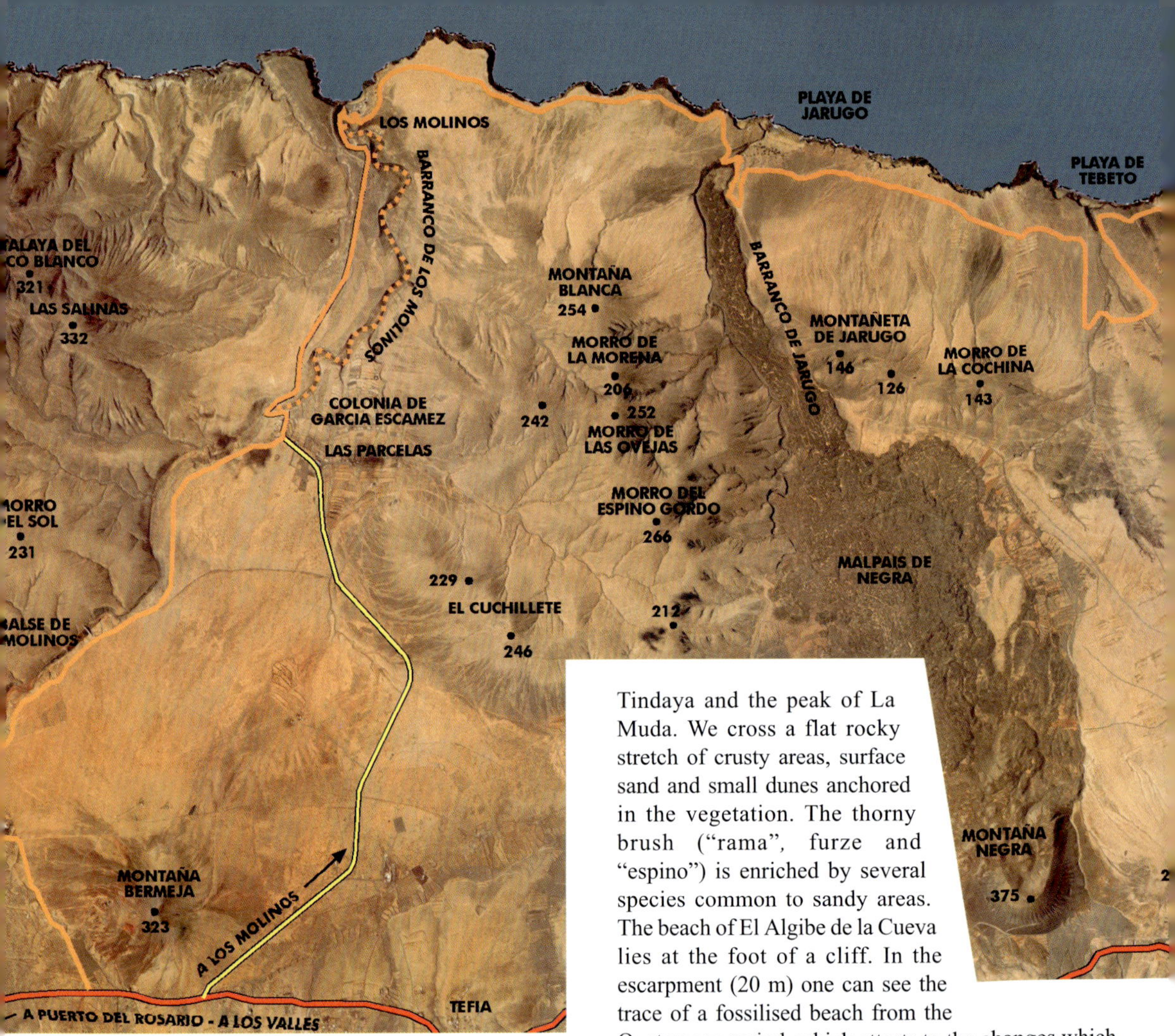

EL COTILLO - LOS MOLINOS

DISTANCE	DURATION	DIFFICULTY
31 km	4-5 hours	Low

Access: From the castle of Río Roque, south of El Cotillo.

Description: The track passes over a *rasa*, a formerly submerged marine platform. At the bottom of the cliff there are a number of golden sandy beaches, some formed around the estuaries of ravines that flow year round. These are superb sites for the observation of various species, especially flatland birds ("engaña", pipit, stone curlew, houbara bustard), and mud-dwellers such as shanks, plover and sandpipers.

The most noticeable volcanic structure at the beginning of the route is Montaña Blanca, and beyond it, in the same chain, you can make out the outline of Tindaya and the peak of La Muda. We cross a flat rocky stretch of crusty areas, surface sand and small dunes anchored in the vegetation. The thorny brush ("rama", furze and "espino") is enriched by several species common to sandy areas. The beach of El Algibe de la Cueva lies at the foot of a cliff. In the escarpment (20 m) one can see the trace of a fossilised beach from the Quaternary period, which attests to the changes which have occurred in sea level. Behind the beach, a chain of "balancones" can be seen over the dunes and, to the south, a small year-round ravine, surrounded by a thicket of "matomoro".

Aguila is the next beach and access is by a set of steps. The track continues along the cliff to the beach and the Esquinzo ravine, on the slopes of which are layers of fossilised sand. The bed of the ravine is covered with a "matomoro" scrub interwoven with plants of treelike appearance through which run rivulets of water fed by natural springs, in some places forming ponds. One of the most outstanding features of this area is a large dune which runs from the beach to the top of the shore-cliff, on the north bank of the ravine.

The trail follows the ravine until turning south across the southern bank and heading towards the edge of the shore-cliff at Punta de Caletones. The path brings us to the centre of Esquinzo, a deteriorated and practically bare malpaís area. A few procumbent plants ("cosco", "corazoncillo") are dispersed amongst the rocks, while some patches of ground are covered in

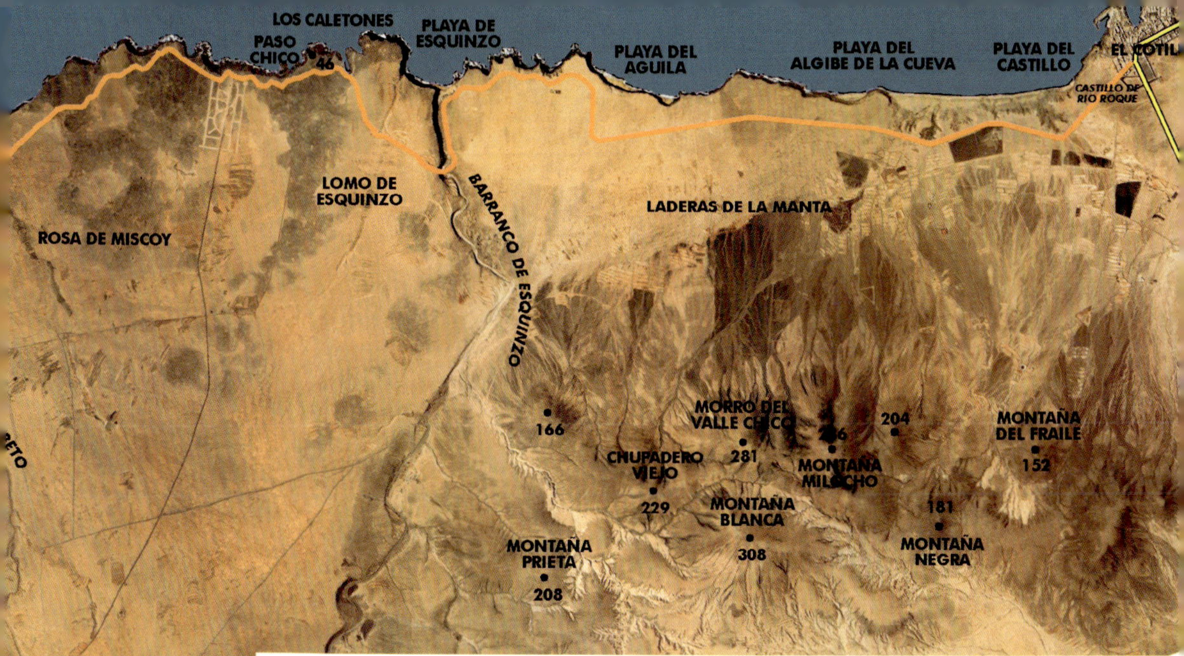

predominantly furze scrub. The colours of the landscape are accentuated with the green, white and ochre tones of the lichens.

Eleven kilometres from the start of the route, we come to a crossroads with the road from Tindaya, and 300 m further on, the geodesic survey point of Paso Chico (40 m above sea level). The track continues to the Tebeto ravine, leaving the malpaís behind for another stretch of rocky flatland. Tebeto has a larger drainage area than the Esquinzo ravine and a similar yellow sandy beach at the base of a cliff, as well as "matomoro" scrub along the banks of the stream. From this point on, the path features landscapes similar to those already seen. The coast is dotted with tiny beaches and natural pools, offering magnificent views from the tops of the cliffs.

Arriving at the Jarugo ravine, we find the last golden sandy beach on this route, flanked by basaltic platforms intensely pounded by the sea. Behind the beach runs a string of dunes anchored by "balancones". The ravine, with running water all year round, is more open than the previous ones, and has the same characteristic scrub. Inland, the peak of La Muda stands out.

To reach the next ravine, Los Molinos, it is necessary to get from the slopes to the level of the river bed via an area where there are some shepherd's huts - this is the only difficult part of the route and it offers breathtaking views. After reaching Casas de los Molinos, the path veers inland. One may decide to end with a walk through the ravine, or follow the paved road to Las Parcelas from where there is a trail leading to the Los Molinos dam (2 km) or the village of Los Llanos de la Concepción (5 km).

WALK (2-3h.): The path follows the bed of the ravine, between the "matamoro" scrub and huge blocks of basalt. At the end of the first curve, at the base of the less steep side, there is an outcrop of fossilised sand. In the escarpment the lighter shades of this formation stand out, trapped between layers of black basalt and red earth. The thorny brush extends to the slopes. The ravine and dam of Los Molinos constitute one of the prime areas of ornithological interest on the island. Flatland birds, mud dwellers and birds of prey are seen along the way, and occasionally one may observe the majestic flight of the Egyptian vulture.

Playa de Jarugo.

PUERTO VENTURA - VILLAVERDE - EL COTILLO

DISTANCE	DURATION	DIFFICULTY
32km	4 hours	Low

Access: From the road connecting Puerto del Rosario and Corralejo there is a sideroad leading inland from Puerto Ventura (Parque Holandés).

Description: The main attraction of this route is that it offers views of the island's main geological formations, from pre-Miocene to the more recent volcanic periods.

The trail leads to the Roja malpaís, which extends outwards from the base of the Roja volcano, the outline of which looms to the right, near the coast. Inland, to the left, the closest peak is Caimán, which with Roja and Escanfraga form a volcanic chain along a fault line running northeast - southwest. These are large cones that did not produce great quantities of lava. The predominant plants in this area are lichens ("aicán", orchilla, *xanthoria sps*.), which add beautiful shades of green, orange and ochre to the colours of the surrounding landscape.

Having partially rounded Caimán, we come closer to Caldera Blanca, Pajarita and Negra, another volcanic series aligned along northeast - southwest faults. Between these peaks and Escanfraga (the highest, at 529 m) lie the Peña Erguida plains, a traditionally agricultural area, in which the lava has been buried under alluvial deposits. We continue over desolate sand covered patches of "henequén", which shape a strange thorny scrubland. We recommend a short walk (30 to 45 min.) across the cultivated areas in the direction of the slopes, which will allow closer observation of pre and post-Hispanic building and irrigation techniques. One may also see the shrike, a small bird of prey which uses the henequén bushes to spear and dismember its victims.

There are now two options: you can continue across the centre of the plain to Villaverde, or leave the main trail and head towards Montaña Pajarita. The second option takes us to the base of Pajarita and Montaña Negra, from where we can take in magnificent views of Peña Erguida, Escanfraga, and Caimán to the south, and Lobos, Lanzarote and the sands of Corralejo to the north. Back again on the main route, we reach Villaverde, and by road, the town of La Oliva, from where we can head onto Lajares. This last stretch of road has a turnoff to the left, along a section of the old main

Montaña Pajarita.

road to the La Arena mountain. This is one of Fuerteventura's more recent volcanoes, with its malpaís of connected lava flows. Together with Los Saltos and Montaña del Mal Nombre it forms a third Quaternary volcanic chain (NE - SW).

The landscape is a gently sloping caliche plain, covered by open spiny scrub ("rama", "matomoro", furze, "espino" and patches of "gramineas" and "cosco"). The palm tree silhouettes stand out amidst the disused cultivation areas, shepherd's buildings and farmhouses. Our walk now leads us away from the remnants of Recent Volcanism to much older geological formations. In the distance we can see a cluster of small pre-Quaternary cones (Prieta, El Fraile, Los Corraletes), among which two

El Jablito.

higher mountains, which are part of the Basal Complex, stand out (Blanca and Milocho). These are among the oldest formations on the island. To the left hovers the peak of Tindaya, Fuerteventura's magic mountain, which is of tremendous archaeological (ruins, artwork) and geological importance (magnificent example of a trachytic formation).

Having passed the palm trees, we come to a fork in the track, and take the route to the left which, after a curve, takes us towards Montaña Blanca. After crossing a ridge, one can see the town of Lajares and the dunes of El Jablito, whose colours offer a fine contrast with the landscape. Small clusters of "tabaiba" on the surrounding slopes add to the composition of the scrub.

Having reached a fossil sand quarry, in an area of extreme beauty, the main track turns to the left; we take another track which goes on in the same direction. This track leaves Montaña Blanca to the left, and to the right, a large white sand dune. The ground becomes increasingly sandy and the scrub is speckled with the species common to such terrain. In this area it is extremely important to stick to the trail.

We recommend a short walk (30 - 45 min.) to the dune, passing through a fossil sand area, through which the rain has carved small gullies. The wind blows away the lighter materials leaving small stones on the surface. A closer examination of these "stones" and their carved contours reveals them to be the fossilised nests of invertebrate ancestors of the wasp. Around the dune itself, the vegetation is typical of sandy areas, and from the top, one can contemplate El Jablito in all its glory.

The trail continues through sandy areas, crossing small ravines, and passing between Los Corraletes and El Fraile, from where we can see from afar the urban centres of El Roque and El Cotillo. In the vicinity of the cones, the terrain is more rugged, apparently covered with a light layer of slack, on which grows a thick furze and "rama" scrub. From here, the track leads to El Roque, coming out onto the road to El Cotillo.

SALINAS DEL CARMEN - POZO NEGRO - CASAS DE JACOMAR

DISTANCE	DURATION	DIFFICULTY
25 km	4-6 hours	Medium to low

Access: Path leading south from Casas de las Salinas, along the top of the cliff.

Description: The saltworks and their associated installations, constitute an important cultural and ethnographic legacy. Birds predominate among the wildlife found in their vicinity. Among them, the limicolous or mud-dwelling birds are the most common, the majority being migratory species. An exception is the blackfooted plover, the only bird of this family which nests in the Canary Islands. Note the subtle differences in size and shape of feet and beaks amongst the avocets, shanks, curlews, sandpipers; a fine example of adaptation for optimum exploitation of available resources.

From the saltworks, the track leads us to a flat rocky zone on top of the escarpment. We travel over a narrow strip of ground from the Second Volcanic Cycle (Basaltic Series 11), between the banks of the La Antigua ravine which nowadays flows into La Torre, and whose estuary includes a pebble beach, a lime kiln, and a tide-washed area covered with mud and an open scrub of "matomoro", tree-like tamarisk and palm trees. The track ascends the southern slope, over land from the island's early sub-aerial lava field (Basaltic Series I), which time has sculpted into its characteristic eroded volcanic appearance.

Two km from the La Torre ravine, we arrive at the Caleta de la Ballena. Between this cove and Caleta Blanca there is an area covered in alluvial deposits from the Majada Honda and Monte Agudo ravines. Access to either cove is difficult. On leaving Caleta Blanca, there is a turn-off to the left, but the main track continues inland across the stony tract of Majada de Cabrones. The scrub, predominantly comprised of furze, is restricted to the water courses between the slopes; and the larger rocks are totally covered in lichens. The track climbs the northern slope of the Majada de Las Cabras ravine, offering magnificent views dominated by Montaña Agudo (494 m). At the top of the slope there is a fork in the trail, and taking the path on the left, you cross to the other side of the ravine and continue in the direction of the coast by way of Cabeza de Leandro.

Access to the Leandro ravine is complicated by the poor condition of the track and the deep potholes. A turnoff in the ravine bed takes us to a beach of pebbles and black sand, with a 2-3 m high platform running along the base of the escarpment forming a natural boardwalk of great beauty.

After climbing the next slope, you can see the settlement and beach of Pozo Negro. Egyptian vultures, "aguilillas" and kestrels can be seen here frequently. We now continue straight along the main path, passing a turnoff to the right. In Pozo Negro the classic lines of the U-shaped valleys were transformed by lava tongues from the Malpaís Grande

Pozo Negro.

(Series IV - Recent Volcanism) which reached the coast, where there is an extensive beach of pebbles and black sand.

From Pozo Negro, the track runs between the southern slope of the valley and the malpaís, along the base of extensive fans of varied alluvial deposits. Some 3.5 km from Pozo Negro, in the vicinity of Loma de la Atalayita, we come across the ruins of aboriginal structures on the malpaís. "Mimos" (an introduced species) have spread along the lava and changed the original landscape.

Passing Las Casas de Saladillo, the route approaches the Grande Valles road, from where one can see the volcanic formation with the four calderas running north-south which formed the Malpaís Grande. Rounding the hill of El Saladillo the trail curves away from the road and heads towards the coast through the El Valle de la Cueva ravine. The track winds along the floor of the U-shape valley between Atalaya de Pozo Negro and the Cuchillo de Juanicón.

Halfway through the valley, the ravine bed begins to be partially covered by malpaís generated by lava from the Jacomar caldera, which spilled over the southern slope and came very close to the coast. This recent volcano (Series IV) surged up among the gorges, discharging most of its lava into the sea, and creating a 2 km long platform at the foot of the cliff. The lava surface is highly irregular with many features formed by gas bubbles in the lava which remained hollow during cooling. These were later used by the first islanders as dwellings or shelters and are now the preserve of such birds as shrikes, the Egyptian vulture, kestrels and "aguilillas".

On the coast, there is a pebble beach with some small dunes of black sand, and an interesting scrub of "uva de mar", "balancones", "matogota" and "barrillera". We recommend a walk (2-3h) along a path which follows the malpaís coastline as far as Caserío de Jacomar. The trail passes by a fossil beach with natural pools and small coves of golden sand, frequented by mud-dwelling birds. We then cross a platform formed by overflow of lava from the Caldera de Jacomar. Inland you will see the cliff which once marked the original coastline. From Jacomar, we return by the same path, or by another that runs along the base of the cliff leading back to the original departure point. From there we pass back through the ravine of Valle de La Cueva to the main road.

THE BETANCURIA MASSIF
(Casas de Mézquez - Ajuí - Aguas Verdes)

DISTANCE	DURATION	DIFFICULTY
34.5 km	5-7 hours	High

Access: In Casas de Mézquez, the paved turnoff on the Ajuí road, leading to the La Peña dam.

Description: This route takes us along the edge of a section of Betancuria Massif, an itinerary that includes the oldest volcanoes on Fuerteventura. At a distance of 4.8 km from Pájara on the road to Ajuí, in the vicinity of Casas de Mézquez, we take the paved turnoff to the right leading to the La Peña dam. We pass over rocky terrain with scrub of "rama", furze and some scattered "tabaiba". The route follows the La Peñita ravine until it joins that of Mal Paso, whose gorge has been carved out of the outer dike of a massive circular plutonic structure. Gulls, tamarisk, reeds, and small groups of Canary palm trees announce the proximity of the Degollada del Mal Paso, flanked by the remains of the huge multi-sectioned dike, and the peak of Aguilillas and the Las Peñas cliff. An unpaved turnoff to the right leads to the latter promontory. Leave the vehicle at the base of the cliff, near the quarries and visit the dam (1-2 h).

We cross a small grove of palms boxed in by the narrow pass and the course of the ravine, and continue along the track on the other side which leads to the chapel of Nuestra Señora de la Peña and the dam. The path and chapel were built with local stone and blend in perfectly with the craggy environs. The ravine is very steep and its bed is clear, with the exception of a few huge rocks fallen from the sides. Ponds have developed in the gullies surrounded by mosses, watercress, wild celery and reeds which are frequented by insects and birds ("alpispas", chats, warblers, "gangas"). The smooth walls are grooved with a complex array of dikes of varying thickness, giving them a striped appearance.

The dam itself is one of the best places on the island for observation of aquatic migratory birds. The community changes over the course of the year, with some species staying longer (ducks, coots), and others seen less frequently. At the other side of the dam there is a small grove of palms near a large house at the foot of El Pico de la Muda, and further on, the Vega de Río Palmas. The entire plutonic circular structure forms a majestic landscape.

We return to the paved road, taking a turnoff to the right which leads us to the track that follows the course of the ravine as far as Ajuí on the coast, through ever thicker clumps of tamarisk and palms. We now find ourselves on a strip of material originating from the volcano of Morro Valdés (Series II) which reaches the sea by way of the ravine of La Peña. This ravine and that of Ajuí include a heavy concentration of dikes cut through greenish, lightly striped rocks, believed to be of undersea sedimentary origin, and more than 100 million years old. From the black sand beach, heading north along the base of the cliff, you can reach the Caves of Ajuí, carved out by the sea, close to a large lime kiln.

Some 200 - 300 m from Ajuí we take a path leading out of the ravine, up the north slope. At the second crossroads, turn left and continue up the slope. Reaching the top, we take a turnoff to the left which leads to the lime kiln, from where there is a fine view of this part of the coast. We leave this crossroads and continue northwards along the top of the cliff.

We now reach the La Peña ravine and following its course among the tamarisk, we come to a pebble beach and the spectacular Arco del Jurao. The dikes are still in evidence, and it is no surprise that this massif is considered to be the largest such dike field in the world.

We return to the path again and 500 m from the beach take a turnoff to the left passing under the tamarisk, close to a line kiln and other constructions. Moving progressively further inland on a ridge between the ravines of La Peña and Sojames, we reach the Tablero de la Peña Vieja at the upper end of Sojames. Here, we find an islet of volcanic materials generated by Morro Valdés (Series II), surrounded by other materials of varied geological origin. Two separate strips can be distinguished at

El Jurao.

this elevation for the remainder of the route. A one kilometre long marine platform, covered in solidified sand, runs parallel to the coast at a height of 20 m; and sloping upward to the middle of La Atalaya (724 m), you will see submarine volcanic lava with its typical cushioned relief.

From Tablero de la Peña Vieja we follow the track to the left, descending into the Tarajalito ravine. Extreme caution is required at this point. Access to the next ravines as far as La Galera is made difficult by the state of the track. After La Galera, we enter the Janey ravine whose mouth includes a fossil beach at an elevation of 10 m, massive dike formations arrayed parallel to the coast, and natural springs frequented by livestock and birds.

The trail improves gradually as it skirts the upper ends of small ravines as far as El Cangrejo. Once here, the Aguas Verdes housing development and the end of the route are visible in the distance. A number of paths lead to the edge of the cliff, and the trail itself leads directly to the tourist complex situated close to the La Galera ravine, with vegetable gardens and a scrub of "matomoros". On the northern slope we see Tablero del Golfete, an island of lava thrown up by Montaña de Tirafe (Series III, Second Volcanic Cycle) which still resists erosion, and stands out amongst the tinges of the materials described above.

THE MALPAÍSES

(Valles de Ortega - Carretera Grandes Valles)

DISTANCE	DURATION	DIFFICULTY
(24+12) km	4-5 hours	Low

Access: The road from Valles de Ortega to Casillas de Morales, at the turnoff for the track to the Gairía caldera (Aguabueyes).

Description: This route has two possible variations: the first takes us around the perimeter of Malpaís Chico, and the second connects with Malpaís Grande. The big old houses and cultivated clayey fields between Valles de Ortega and Casillas de Morales are a reminder of the days when agriculture was much more important to the region. On reaching Casillas de Morales, the paved road ends and a track continues southward in the direction of the Aguabueyes volcano (416 m) and the Caldera of Gairía, a cone open to the southeast, from which flowed the lava forming the Malpaís Chico (10 km2). To the left, like islets, rise geological reminders of the Ancient Series (Montaña de Parrado, Montaña de la Torrecilla) that were engulfed by lava. In the background loom the series of calderas which formed Malpaís Grande, and beyond them, the *cuchillos* (knife-like eroded ridges) of the Ancient Series, which lie perpendicular to the coast and separate great U-shaped valleys (Pozo Negro, Gran Tarajal).

The track advances through caliche and clayey terrain. After passing a turnoff to the right some two km out, we reach Malpaís Chico, a garden amidst the lava, home to one of the island's most outstanding "tabaiba" fields. At the top of Montaña del Parrado, we pass a turnoff to the left and round Gairía by its base of "tabaiba dulce" covered slopes. We leave the malpaís and pass over more caliche strewn ground, climbing upwards to where Tiscamanita and Tuineje can be seen in the distance. We cross a small valley with terraced *gavias* and climb the next slope where, having passed a lime kiln on the left, we reach a track coming from Tiscamanita (7.5 km from the start). From this point, the route offers two alternatives.

ALTERNATIVE 1. We take the trail from Tiscamanita and head in the direction of Caldera de la Laguna, a rocky cone of volcanic slack which bears the marks of quarrying. We pass along a caliche and clayey strip that separates the two malpaíses.

At 2.5 km from the fork, there is a turnoff to the right that leads to the quarries, from where there is a magnificent view of the Malpaís Chico and part of Malpaís Grande. We return to the main track which continues in the direction of Los Ralcones, a geological reminder of the Ancient Series. We pass by a well surrounded by poplars and come to the end of this tongue of ancient materials. We are now in the Laguna de las Mujeres, a place where the lava flows of the malpaíses Grande and Chico met. The trail crosses a small tongue of Malpaís Chico, with a grove of "tabaiba", and at the next crossroads we turn left.

We pass over clayey ground, occasionally strewn with rocks fallen from the slopes, between the Malpaís Chico and the eroded remnants of the Ancient Series formation. At the top of La Atalayita, we come to a crossroads. The track to the right connects with the roads to Puerto del Rosario, Antigua and Pozo Negro. The track to the left takes us around the edge of the malpaís via the base of Montaña de La Goma to a paved road. After three km, we reach Valles de Ortega, and have returned to the starting point.

ALTERNATIVE 2. As mentioned earlier, both variations of the route begin in the same place, which we can reach either via Casillas de Morales (7.5 km) or directly from Tiscamanita (1.5 km). We leave

Caldera de la Laguna, a source of volcanic ash.

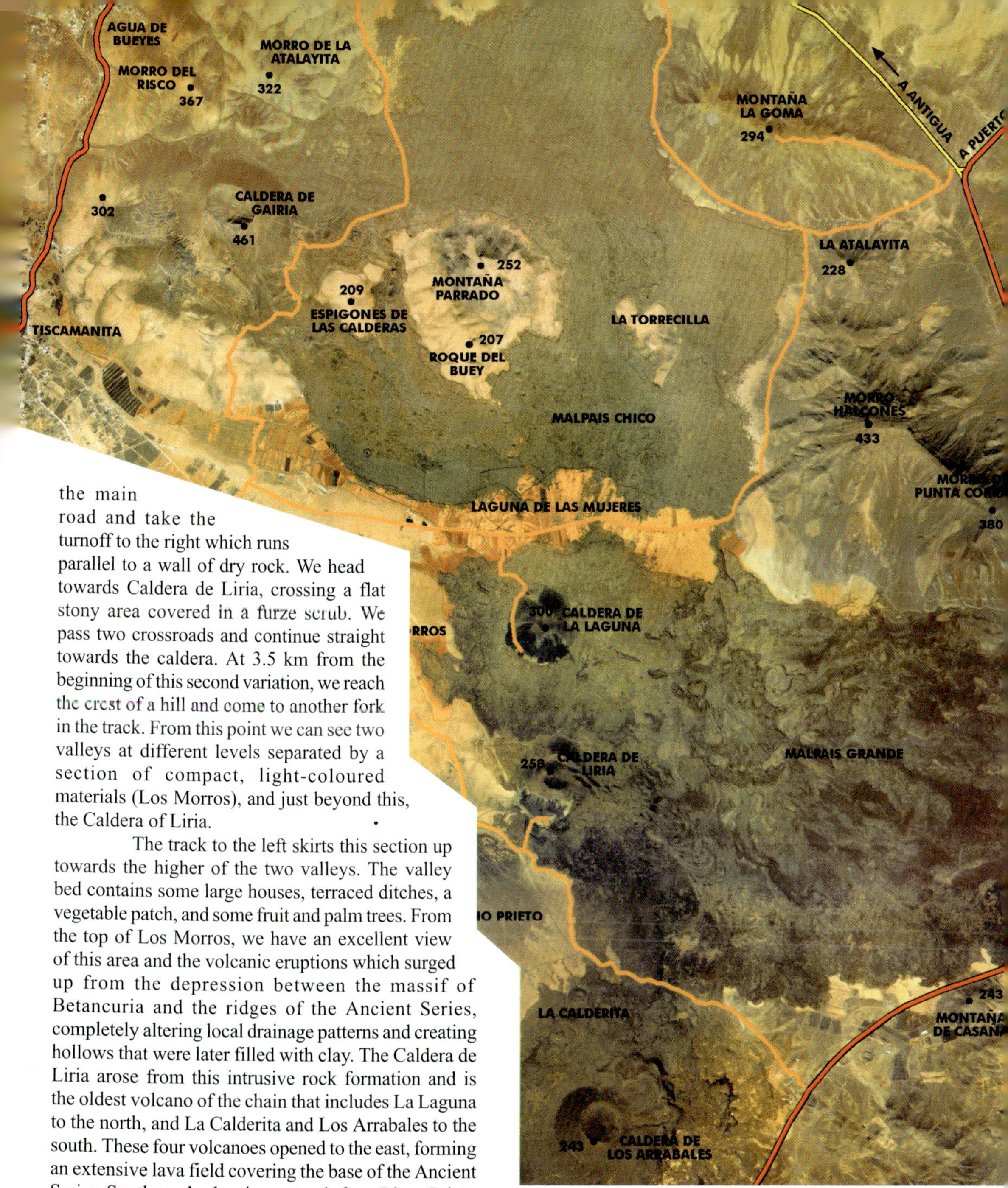

the main road and take the turnoff to the right which runs parallel to a wall of dry rock. We head towards Caldera de Liria, crossing a flat stony area covered in a furze scrub. We pass two crossroads and continue straight towards the caldera. At 3.5 km from the beginning of this second variation, we reach the crest of a hill and come to another fork in the track. From this point we can see two valleys at different levels separated by a section of compact, light-coloured materials (Los Morros), and just beyond this, the Caldera of Liria.

The track to the left skirts this section up towards the higher of the two valleys. The valley bed contains some large houses, terraced ditches, a vegetable patch, and some fruit and palm trees. From the top of Los Morros, we have an excellent view of this area and the volcanic eruptions which surged up from the depression between the massif of Betancuria and the ridges of the Ancient Series, completely altering local drainage patterns and creating hollows that were later filled with clay. The Caldera de Liria arose from this intrusive rock formation and is the oldest volcano of the chain that includes La Laguna to the north, and La Calderita and Los Arrabales to the south. These four volcanoes opened to the east, forming an extensive lava field covering the base of the Ancient Series. Southwards, the view extends from Llano Prieto as far as Gran Tarajal and the sea. To the north is Malpaís Chico and Gairía, and to the east, beyond Malpaís Grande, lie the crags of the Ancient Series.

We return to the track and descend to the Llanos de Prieto, a plain bordering Los Morros. The surrounding slopes are covered in furze and "rama", and we pass by the foot of quarry sites on Caldera de Liria. Once on the plain, we take a turnoff to the left that leads to the malpaís in the direction of La Calderita and Los Arrabales. The scrub on Malpaís Grande is distinct from that of Chico, "rama" dominating to the detriment of "tabaiba", and there being a markedly greater presence of furze, "espino" and "mimos". After passing the base of the north slope of La Calderita, we cross islets of ancient terrain, and come out at the road to Grandes Valles between Los Arrabales and Cuchillo de Los Olivos.

ISTMO DE LA PARED

(La Pared Isthmus) (La Pared - Pecenescal)

DISTANCE	TIME	DIFFICULTY
19 km	4-5 hours	Medium (4x4)

Access: Turn off the isthmus road at the La Pared housing development.

Description: The La Pared isthmus connects the Jandía peninsula to the rest of Fuerteventura, through the island's most extensive *jable* (dune field), it also being one of the areas inhabited by the treasure of its fauna, the houbara bustard. The mountains hovering to the north are the extremity of the Betancuria chain, where the Basal Complex comes to the surface. To the south, the sands sweep over the Ancient Series basaltic formations as far as the ravine of Pecenescal, beyond which rise the *cuchillos* of the Jandía Massif. We can see an old jable made up of caliche and solidified sand and modern jable comprising longitudinal dunes blown across the isthmus by the trade winds from the west (windward) to the east (leeward) coast. The mobility of the marine sands is the defining characteristic of this eternally changing landscape.

The contrast between the black sand of La Pared beach at the outlet of a ravine, with the golden sand of Viejo Rey beach at the foot of the cliff, reflects different geomorphologic processes. The southern coast is marked by a 20 m escarpment comprising gentle slopes which are cross-sectioned by short canyons. In the jable, it is best to avoid the labyrinth of tracks bordering the cliffs, taking instead the most clearly marked one which winds behind the inner tips of the ravines. Following the other routes entails the possibility of harming this magnificent landscape, not to mention the risk of being trapped.

At the beginning of the route, we travel over smooth, occasionally rocky slopes, with a surface covering of sand, and small dunes anchored by plants. As we continue, the proportion of sand on the ground increases while the elements of thorny scrub ("rama", "espino", furze) diminish and are replaced by sand-loving species ("ononis", "uva de mar", "lecheruela", "algahuera", "gamonilla", "lengua pájaro", "corazoncillos"). Leaving Viejo Rey beach behind, with a lime kiln at the top of the cliff, we pass by some apartments and take the track that runs about 300 m inland from the cliff.

From the crest of a hillock, we have a fine view of the coast and jable. In the distance, the profile of a perfect cone (Montaña Aguda) stands out on the Jandía Massif, and to the left, the peak of La Zarza (807 m), the highest point of the island. Amid the sands we see Loma de la Ruda, a promontory scarred with trails

and separated from the massif by the Degollada de los Mojones. This is one of the few islets of ancient materials to emerge from the sands, while the closest heights (Alto de Agua Oveja, Granillo) are fossil dunes, with a surface covering of caliche.

At about 5 km from the start of the route there is a turnoff to the right which descends to the Cañada de Agua Liques. On the coast, the cliff face reveals alternating layers of basalt and fossilised sand, in which the combined erosive action of wind and sea has sculpted remarkably beautiful forms. At the base of the escarpment there are natural springs frequented by birds and livestock. The availability of fresh water and shellfish, especially mussels, meant that this spot was attractive from the earliest times, as attested by the number and size of shell beds. A marine platform runs along the base of the cliff forming a natural boardwalk which adds to the area's charm.

Returning to the main route and passing Alto de Agua Oveja, we take another turnoff to the right leading to Veril Manso, a point further south similar in nature to Agua Liques. The most outstanding feature are the huge dunes that stretch upwards from the beach to the cliff top. We return again to the main route, crossing the Loma de la Ruda (9.5 km from the start) which is covered with a scrub of "ononis". We leave this very old formation behind and enter the Degollada de Mojones, an area of almost bare sand that stretches

to the first promontories of the Jandía Massif, where such species as "tabaiba dulce" begin to enrich the vegetation.

The foot of El Cuchillete offers a good view of the massif and the isthmus. On the shores of the windward Barlovento beach, huge dunes ascend to the cliff top. A track enables us to go down further, to where we find small canyons hollowed out from the fossil sand. The beach can be reached via the dunes.

We leave the windward slope, via the Degollada de Los Mojones to the Pecenescal ravine. The almost bare sand gives way, in the upper part of the Degollada, to a scrub of "ononis" and furze and some isolated wild "tabaiba". The route takes us over slopes made up of very old material and sand-filled stream beds, to the slopes of the massif and a contour of fossilised sand four to five metres high.

Close to Loma Negra (322 m), an islet of old terrain protruding from the sand, we come to a fork, the right branch of which skirts the base of the massif. We continue straight ahead along the ravine bed, passing between contours of solidified sand.

The route ends at km 34-35 of the road to Morro Jable, where it meets the Pecenescal ravine at the foot of Risco del Paso (253 m), another ancient formation. On the other side of the main road, a trail continues to the Sotavento beach, with its large dunes, ponds and the best salt marshes of the Canary Islands.

Barlovento de Jandía.

PICO DE LA ZARZA

(Jandía Massif).

DISTANCE	DURATION	DIFFICULTY
15 km	5-6 hours	Medium-High

Access: The road running from the west of Morro Jable to Valle de Vinamar.

Description: From a number of viewpoints, Jandía can be considered a separate entity from the rest of Fuerteventura. Its use as pasture land goes back to the time of the aborigines, and more recently the peninsula has been used for agricultural purposes. Today, tourism is the only economic activity of any importance. The jable of the La Pared isthmus gives way to an ancient massif (11 million years old, Series 1), without recent (Quaternary) eruptions, and highly eroded. Jandía possesses the longest beaches, the highest elevations and the greatest botanic wealth of the island.

This route is classified as of medium-high difficulty because it must be done on foot, and includes a climb of 800 m over a distance of 7 km in the semi-desert conditions which typify the island. Good footwear is essential, as is adequate sun protection for the head and skin, not to mention water and some food.

The route begins on a path at the entrance to Morro Jable (at the RIU Apartments). There are

Cardón canario.

two options for the initial climb: the first is steep at the beginning, taking us to the crest of Tabla de Vinamar, between the ravines of Vinamar and Butihondo; the second follows the bed of the Vinamar ravine to the base of Degollada de Hierba Huerto, from where we ascend along the east slope to the crest of Tabla de Vinamar.

We walk through thorny scrub with scattered patches of thistle ("cardón canario"). On the higher slopes, the "tabaiba" occupies the steepest spots offering any foothold. The rocky escarpments appear to be tinged with the orange and green lichens.

From the crest of Tabla de Vinamar (550 m), we continue to the Pico de la Zarza (807 m), which offers stupendous views.

View of Punta de Jandía from Pico de la Zarza.

Erosion has carved out deep ravines on the southern flank, whose contours are readjusted by occasional landslides. These ravines spread outwards from an arched dividing line, cutting up the old formation into narrow elongated ridges.

In the north, the present-day coastline is abrupt, with a jagged relief which tells us that originally it extended considerably further. The slopes tend to be shorter and steeper than in the south. The vegetation is found on elevated strips of land, while on the coast one can make out the sandy areas and dunes anchored by "balancones"; then behind them thorny scrub turning to thistle halfway up the slopes and finally, near the summits, some traces of mountain verdure.

The existence of the latter is thanks to the humid trade winds, which are only blocked by the highest peaks on the windward side. The low number of plants is largely a consequence of clearing and overgrazing. The plant life is now being recovered.

In addition to the arboreal elements of mountain vegetation (wild olive tree, "peralillo",) the botanic treasure of the summits of Jandía includes a number of species which are endemic to Fuerteventura ("magarza", "tajinaste de Jandía"), the Canary Islands ("Sideritis pumila", "Reichardia famarae", "Monanthes microbotrys"), and the Macaronesian Archipelago ("tasaigo", "morgallana", "Bavalia cariensis" and "Cheilantes maderensis"), and also includes ferns and minuscule plants growing in the crevices of rocks. One very abundant endemic plant is "jorao", with large yellow flowers, whose scent lingers around the summits, becoming one of the main features of the climb. Even amongst the lichens there are a number of endemic species, the most noteworthy being orchilla, (used as a source of dye in the past), which grow on rock faces exposed to the sea breeze.

At more than 800 m, the Pico de la Zarza gives us a view only matched by that of the birds of prey (Egyptian vulture, "aguilillas", crows, kestrels) circling through the warm air currents which they use to move from one side to another. To the northeast loom the last spurs of the Jandía massif, the La Pared dune fields, the Betancuria massif and the *cuchillos* of Grandes Valles. To the east, rise El Fraile, Licanejos Sierras, Montaña Aguda and Montaña Azufra in the Sierra of Valluelos; and the El Cotillo plain and the lighthouse. To the south, the valleys of Butihondo, Vinámar and El Ciervo which flow into Morro Jable, the salt marsh of El Matorral, and the large Sotavento Beach. Finally, on the north coast, are the beaches of Cofete and Barlovento, separated by the islet of Jandía, a circular dike joined to the coast by a tiny isthmus of sand.

In the solitude of these mountains, valleys, and beaches, there still roam wild goats and donkeys, a reminder of just how special Jandía is.

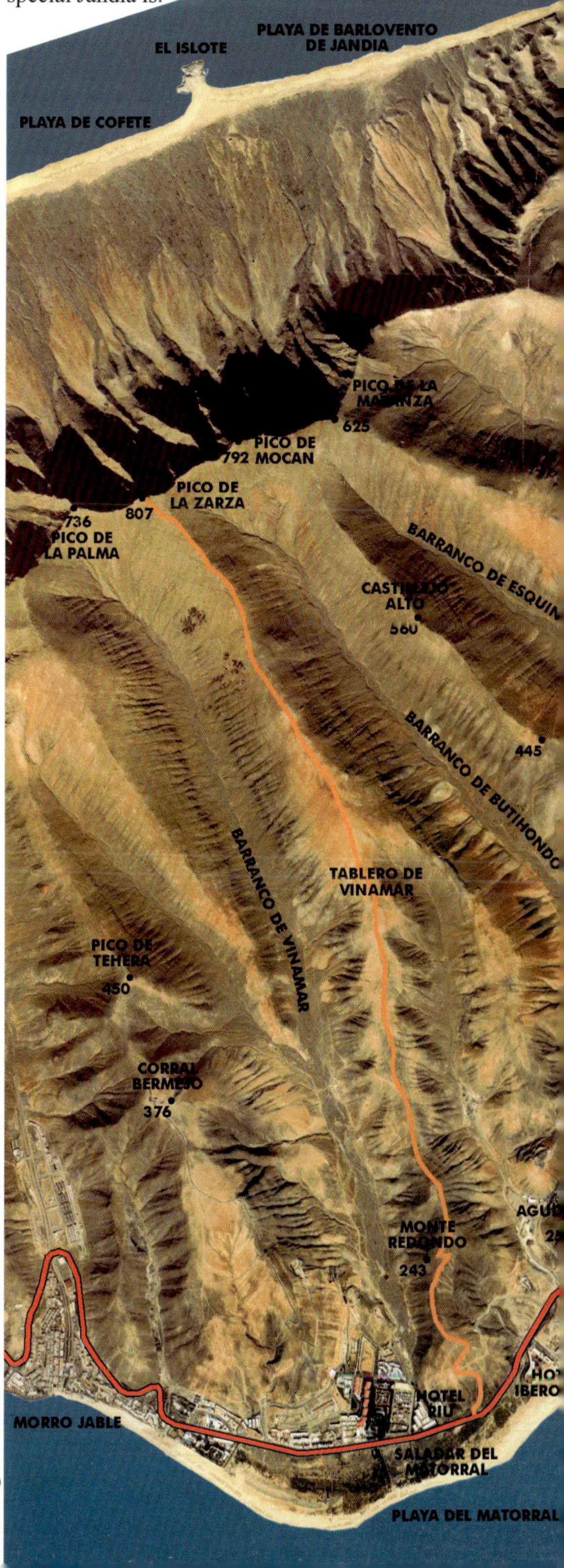

THE JANDÍA MASSIF
(Morro Jable - Faro de Jandía/Cofete)

DISTANCE	DURATION	DIFFICULTY
80 km	7-8 hours	Low

Access: Turnoff from the road to the Morro Jable port.

Description: This route takes us around the very old (Series 1) Jandía massif, stripped bare by erosion, which reaches an altitude of 807 m at the Pico de la Zarza, the highest point on the island. We leave the road to Morro Jable port on a turnoff to the right that runs along the top of the cliff. The landscape is characterised by a thorny scrub ("rama", furze, "espino"), and small caliche-covered rocky slopes, giving way to wide plains. Leaving the slopes of Cuchillo del Ciervo, we enter Gran Valle, crowned by the Degollada de Cofete and Pico del Fraile (683 m). On the west slope we see the Morro de Mungía, an area of archaeological importance close to Siete Fuentes.

About 6.5 km from the start we reach Caserío de Jorós, the surroundings of which are dotted with greenhouses, and scattered tamarisks, reeds and palm trees. Leaving the Jorós valley, we pass through that of Los Escobones and arrive at the valley of Los Mosquitos, between Cuchillo del Palo to the east and Sierra de Licanejos to the west. On the slopes and floor of the valley, around the road, we find an extensive population of "cardón de Jandía" (Jandía thistle), a species not found anywhere else in the world. This extremely rare and beautiful cactus-like plant with its huge thorns is now a protected species.

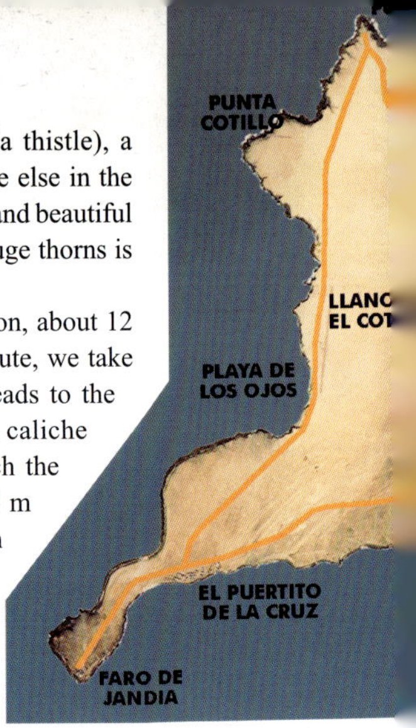

One km further on, about 12 km from the start of the route, we take a turnoff to the left that leads to the coast over a slope of old caliche encrusted ground. We reach the Juan Gómez beach, a 200 m stretch of golden sand, with generally little wind and a calm sea. We return to the main road and continue westward for about 300 m to where a cross marks a fork leading either to the Jandía lighthouse or Cofete.

ALTERNATIVE 1: Taking the left fork we leave the Valle de Juan Gómez in the direction of the lighthouse and village, where a modern windmill stands out on the horizon. We travel along the base of Sierra de Valluelos to the Llanos del Cotillo, a plain that stretches between Montaña Azufra (to the extreme south of the massif) and Punta de Jandía. The rocky plain is partially covered in sand which forms small dunes anchored by plants. The make-up of the scrub changes, with fewer "rama" and furze in favour of "matomoro" and "uva de mar".

Cofete beach.

We reach a paved road that passes alongside the village and ends at the lighthouse (10 km from the fork). We find ourselves on the extreme southwest of the island, where the north and south seas meet, on a cliff 8-10 metres above the basaltic platform and large dikes on which the waves break.

We return to the village and take a paved turnoff to the left which runs along the west coast of the peninsula. Two km further on, we come to the beach of Los Ojos, a yellow sand beach at the base of a cliff with cross-sectional dikes running into the sea. Six km from the village, the paved road ends at Punta Pesebre (radio beacon), the extreme northwest of the island. We continue on a road that runs along the north coast towards the massif, as far as Caleta de la Madera, a circular inlet open to the north with an escarpment of some 50 m marked with peaks and narrow basaltic tongues. The dark reddish tones of the cove contrast with the luminosity of the sand which rises to the slopes of Las Talahijas. From this point we retrace our path back to Punta Pesebre, the village and the fork in the road.

ALTERNATIVE 2: We take the right fork which climbs the east bank of the Agua Oveja ravine. The slopes are tinted with shades of orange and green from the lichens ("Xanthoria" and "aicán") that cover the rocky scarps and the stony alluvial deposits. About 2 km from the fork, we reach the Degollada de Agua Oveja (220 m), on the axis dividing the peninsula's summits, between Montaña Aguda and the Sierra de Valluelos to the west, and the Sierra de Licanejos to the east. We now begin to descend the north slope of the massif, where we have a fine view of the beaches of Cofete and Barlovento, separated by the islet. Some two km from the Degollada, and at an elevation of 100 m, we reach a thistle field. The "cardón canario" (Canary Island thistle) is also endemic, more widely found here than in Jandía, and larger in size (2 m high, and up to 8 m in diameter); it has a candelabra shape, a more intense shade of green, and much shorter needles. The vegetation in the cactus field is much richer, with species such as the "espina blanca", "tasigo" or "cornical", only found within the thistle clumps where the grazing animals cannot reach them. Above the thistle field, at the foot of the El Fraile cliff, the presence of springs is marked by the occasional palm tree and greener vegetation and flowers.

After rounding a bend, we see the Winters' house below the Degollada de Cofete, and we come to the village. The route continues along the hillside, passing the Winters house and ending at the Cortijo de Cofete, at 1.5 km. From the village there is a turnoff to the coast, which takes us through some thorny brush to Cofete beach (2 km). Upon reaching the beach, with huge dunes anchored by "balancones" and some scattered "uva de mar", "lecheruela" and "lengua pájaro", we recommend a walk of some three km along the seashore as far as the islet, which could be extended to more than 11 km between Punta Paloma to the north and Roque del Moro to the south. The large shellbed at the top of the islet attests to the importance of this resource-rich coast for the islanders.

Returning by the same route, we end at Morro Jable.

ISLA DE LOBOS (LOBOS ISLAND)

DISTANCE	DURATION	DIFFICULTY
7 km	6 h. 30 m	Low

Access: From the port of Corralejo, a glass bottom boat leaves for Isla de Lobos every morning at 10 am, returning at 4 pm.

Description: Lobos is a tiny island of 6 km2, situated in the La Bocaina strait that separates Fuerteventura and Lanzarote. According to dating of fossil remains from an elevated beach (1-2 m), the island is more than 6,000 - 8,000 years old. Until 1968, it was inhabited only by a lighthouse keeper and his family. Such isolation assured the conservation of one of the best sea and land sites in the Canary Islands. The island is now protected by law and must be treated with utmost care and respect by visitors. Essentially, it is a botanic garden set on volcanic terrain with fascinating birdlife and surrounded by extraordinarily rich seas.

A visit to Lobos Island requires previous planning. There is no accommodation, and for camping a permit must be obtained from the town hall in La Oliva. One can take food and water along or else make use of the El Puertito bar, making arrangements prior to arrival on the island. Scrupulous care must be taken with litter, which must of course be taken off the island. Respect the natural resources of Lobos at all times by keeping to the indicated trails. The taking of samples or species is prohibited. Only rod fishing is permitted in designated areas and the catch is limited. Finally, great care should be taken not to disturb the birdlife, especially during the nesting season.

We leave the port of Corralejo, crossing El Río, the stretch of water that separates Lobos from Fuerteventura, which does not exceed 10 m in depth at any point. Geologically, Lobos consists of a recent volcano (Series IV) and the material thrown up by it. La Caldera (110 m) is the end of a long volcanic chain running northeast-southwest on the north coast of Fuerteventura, including some of the volcanoes that produced the Malpaís de la Caldera Encantada (or Bayuyo). Beyond the dunes of Corralejo and the malpaís, we see the contours of Montaña Colorada, Calderón Rondo, Caldera Encantada and Bayuyo, all produced by the same eruptions that formed Lobos.

Disembarking, we take the trail that leads to the east coast, in the direction of Casas del Puertito, to continue around the island. We find ourselves in a slag-filled basaltic lava "malpaís" of very irregular relief. Beautiful "tabaiba dulce" is found on the lava while the rocks have a covering of a variety of lichens dominated by the green tones of "aicán". In addition to the "tabaibas", some of which are extremely large (up to 2 m high), there are several species of bush including "espino", furze, and "matomoros". Approaching the village, we come across several shallow coves with white sandy bottoms; they are used for swimming and mooring fishing boats.

We leave Casas del Puertito and arrive at La Laguna, a clay-filled depression separated from the sea by a strand of pebbles. Some of the species found here (the blue-flowered "siempreviva" and many "matos"), are adapted to prolonged periods of tidal immersion. Around the depression and its patches of the above mentioned species, other plants typically found in salt water areas can also be seen ("uva de mar", "balancones", and a pink-flowered "siempreviva").

La Laguna and other similar depressions are frequently visited by a variety of mud-dwelling birds on their migratory journeys, and this is what makes them sensitive to human activity. Plover, shanks, egrets, curlews, etc., use the island as a stopover for rest and feeding; however, other species of non-limicolous aquatic birds (tern, gull, puffin) also nest and breed on this island.

On leaving La Laguna, large numbers of "hornitos" begin to appear; these are protuberances formed by gases bubbling up through the lava. Some lava areas are covered with marine sand and these gradually grow in size until forming a small dune field on the northeast tip of the island. The sandy areas feature a scrub of "balancones", "algahuera", "uva de mar", "camellera", and "corazoncillos", among others.

From the lighthouse, there is a fine view of the south of Lanzarote, with Playa Blanca beach, the plains of El Rubicón, and the Los Ajaches massif. We leave the lighthouse and pass the remains of a volcanic formation smaller than La Caldera, which has now been almost destroyed by the sea. It is a volcanic cone, comprising layers of stone, solidified bubbles and slag. The slopes are covered with "tabaiba dulce", and harbour a large colony of seagulls and the beautiful predator, the "guincho" or osprey. This area is known as "la cocina" (the kitchen), due to the vapour-like mists ascending the slopes at sunset.

We approach the southwest coast of the island and some old salt fields. The malpaís terrain gives way to a saline clay-covered depression (similar to those in the vicinity of the lighthouse and Las Casas). On occasion, the salt water emerges from the bed. Depending on the degree of salinity, these depressions may either be surrounded by a thorny scrub or covered, as in the present case, in "siemprevivas" 60-80 cm high, which produce pink flowers every spring.

Leaving this area, we find ourselves once again in a dune field, which announces the proximity of La Caleta, a magnificent beach of black sand and shallow water. At the right time of the year, you can both have a swim and observe the terns on Punta del Marrajo. From here we return to the starting point of the route, sailing back to Corralejo, and leaving behind the natural treasure that is Lobos island.

PUNTA MARTIÑO
FARO
CALETA
DEL VINO
PLAYA DE
LA ARENA
MORRO DE FELIPE
CALETA
DEL PALO
MORRO
COLORADO
LA CALDERA
LOBOS
127
MORRO DE
LA MORADA
ROQUE
DEL ESTE
LAS TRES
HERMANAS
LA CALETA
CALETA DE
LA RASCA
PUNTA DEL
MARRAJO
CASAS DEL
PUERTITO
MUELLE

THE WINDMILL ROUTE

Scattered throughout the landscape, the windmills are a reflection of a long period in Fuerteventura's economic history.

Two basic factors led to the building of the windmills: one of which is climatic, the other socio-economic. On the one hand, there are the trade winds, which blow all year round and are especially strong in July and August, just after the harvest. On the other, the island's economy was historically based on cereals.

The combination of these factors in the 17th and 18th centuries led to replacement of the more primitive "tahonas" by windmills. Later, in the 19th century a second type of windmill appeared.

We can distinguish two different types of mills: The "molino macho" (male mill), a design imported from the plains of Castile, has a circular base and truncated cone shape. The milling mechanism was situated at the top and covered by a "caseta" (literally "hut") of wood and sheet metal. The angle of the sails could be adjusted by means of a rudder or axle, in accordance with the direction of the wind. There were usually from four to six sails and two floors, one for the millstones, the other for handling the grain. The mills were sited in prominent areas exposed to the prevailing winds. Made of stone and mortar or lime, they were often finished off with a decorative circular platform built of stone from the surrounding area.

"Molino" at Llanos de la Concepción.

"La molina" (female mill), the more modern 19th century type, has a different design and milling mechanism. Of four or six sails, these mills have a horizontal axle supported by wooden rollers transferring motion through a wheel to a metal shaft whose vertical movement drives the two millstones located in the triangular wooden base. The entire mechanism is supported on a lower axle and a

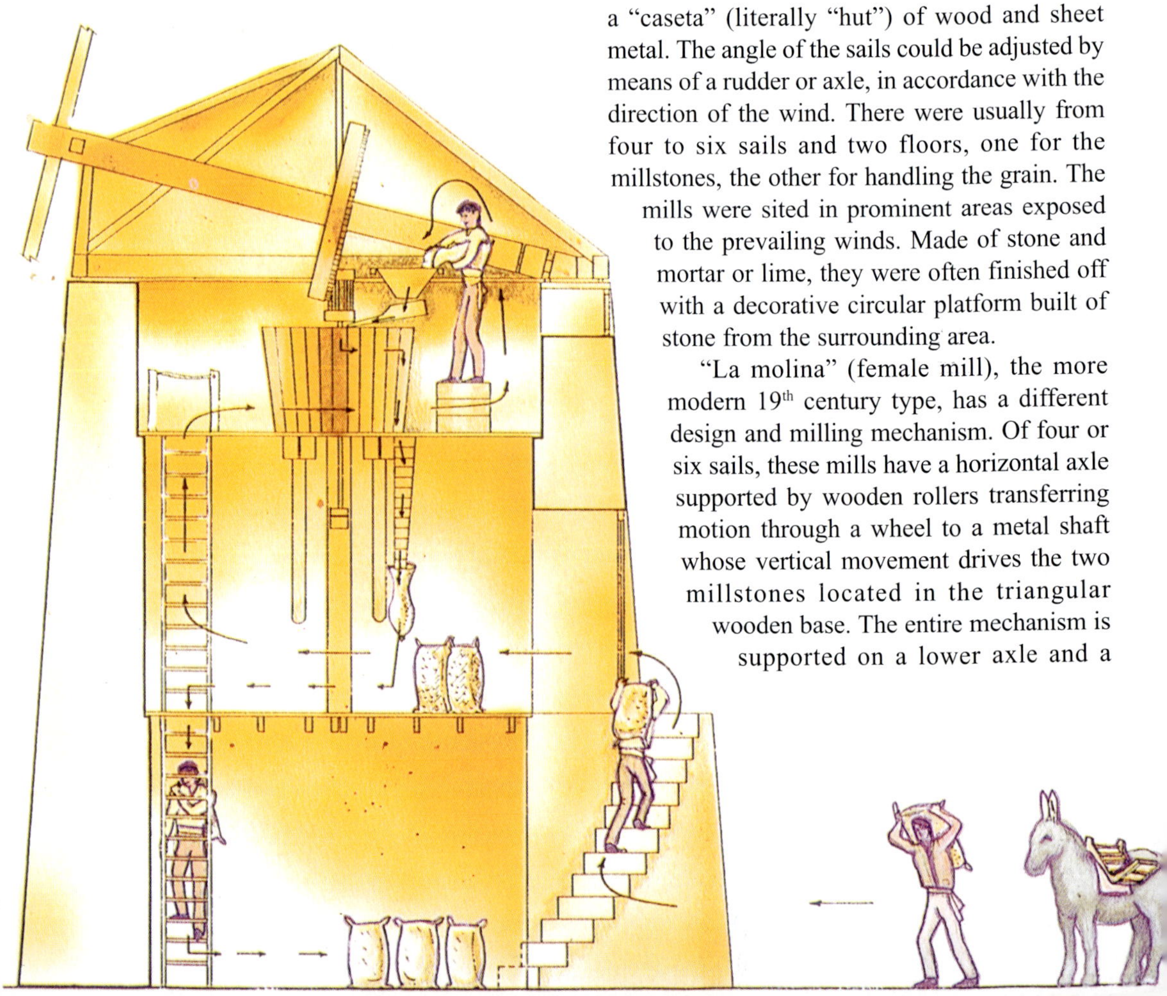

wooden shaft enables it to be oriented according to the direction of the prevailing wind. The mill building itself was made from local stone and mortar, with a quadrangular base, and had either a single room for milling or adjoining rooms for storage and living quarters. There are also some technological curiosities, such as the Molina de Valles de Ortega, which moves two sets of millstones by means of a single horizontal mechanism.

Of the 38 existing windmills, 23 are molinos and 15 are molinas - in various states of preservation. Those belonging to the Island Council are the best preserved, in some cases restored and extended, as is the case with the Molino de Antigua (bar, restaurant, gardens, etc.) Most of the others are quite dilapidated, except those of Lajares and Durazno in Antigua, which have been restored by their owners.

Deserving of special mention is the mill at Tiscamanita, which together with the adjoining house forms a beautifully harmonious structure. It is a traditional two-storey, U-shaped construction around a central patio. The patio features an exterior staircase leading to the upper floor. It is finished in stone and has been preserved in very good condition.

"Molino" at Valles de Ortega.

"Molina" at Puerto Lajas.

Our route includes 13 of the best preserved mills, all on the main road except for Molina de Puerto Lajas (about 6 km from Puerto del Rosario), which is some distance off the route, but is included because it is the only mill that has been restored to perfect condition; the Correlejo windmill is in extremely poor condition, but was also included because it makes for a good beginning and end to the tour.

The mills by order of appearance are the following:

1. Molina de Corralejo (town centre).
2. Molino de Corralejo (town centre).
3. Molino de El Roque.

4-5. Molinos de Villaverde.

6. Molino de Tefía.

7-8. Molinos de Los Llanos de la Concepción.

9. Molino de Antigua.

10-11. Molinos de Valles de Ortega.

12. Molino y Casa de Tiscamanita.

13. Molino de Puerto Lajas.